3 4

WATER RESOURCE PLANNING, DEVELOPMENT AND MANAGEMENT SERIES

SAFEGUARDING THE NATION'S DRINKING WATER

WATER RESOURCE PLANNING, DEVELOPMENT AND MANAGEMENT SERIES

Planning and Managing of Water Resources
Peter A. Unwal (Editor)
2006. ISBN: 1-59454-757-2

Water Resources Research Progress
Liam N. Robinson (Editor)
2008. ISBN: 1-60021-973-x

Trends in Water Resources Research
Alan B. Prescott and Thomas U. Barkely (Editors)
2008. ISBN: 978-1-60456-038-1

Integrated Water Resource Management in the Kurdistan Region
Almas Heshmati
2009. ISBN: 978-1-60741-295-3

Water Quality: Physical, Chemical and Biological Characteristics
Kudret Ertuð and Ilker Mirza (Editors)
2009. ISBN: 978-1-60741-633-3

Water Infrastructure
Harald Nystroem (Editor)
2010. ISBN: 978-1-60692-479-2

Safeguarding the Nation's Drinking Water
Juan A. Schrock (Editor)
2010. ISBN: 978-1-60741-234-2

WATER RESOURCE PLANNING, DEVELOPMENT AND MANAGEMENT SERIES

SAFEGUARDING THE NATION'S DRINKING WATER

JUAN A. SCHROCK
EDITOR

Nova Science Publishers, Inc.
New York

For permission to use material from this book please contact us:
Telephone 631-231-7269; Fax 631-231-8175
Web Site: http://www.novapublishers.com

Library of Congress Cataloging-in-Publication Data

Safeguarding the nation's drinking water / Juan A. Schrock [editor].
p. cm.
Includes index.
ISBN 978-1-60741-234-2 (softcover)
1. Drinking water--Law and legislation--United States. 2. Water--Pollution--Law and legislation--United States. 3. Drinking water--United States. I. Schrock, Juan A.
KF3794.S28 2009
344.7304'6343--dc22

2009029202

Published by Nova Science Publishers, Inc. ✦ New York

CONTENTS

PREFACE

The events of September 11, 2001 focused heightened attention on the security status of the nation's drinking water supplies and the vulnerability of this critical infrastructure sector to attack. As discussed in this book, Congress since has enacted security requirements for public water systems and has provided funding for vulnerability assessments, emergency planning, and drinking water research. The Environmental Protection Agency (EPA), the lead federal agency for the water sector, has worked with water utilities, state and local governments, and federal agencies to improve the drinking water security. Thus, although EPA, states, localities, and water utilities have taken steps to address security concerns, the security of the nation's water supplies continues to attract congressional attention. This book reviews governmental and water utility efforts to improve drinking water security.

Chapter 1 - Much progress has been made in assuring the quality of public water supplies since the Safe Drinking Water Act (SDWA) was first enacted in 1974. Public water systems must meet extensive regulations, and public water system management has become a much more complex and professional endeavor. The Environmental Protection Agency (EPA) has regulated some 91 drinking water contaminants, and more regulations are pending. In 2005, EPA reported that the number of systems reporting no violations of drinking water standards reached a new high of 94% in 2003. Despite such progress, however, an array of issues and challenges remain.

In the 110th Congress, key issues have involved infrastructure funding needs, related compliance issues, and concerns caused by detections of unregulated contaminants in drinking water, such as perchlorate and pharmaceuticals and personal care products (PPCPs). Another issue involves the adequacy of existing regulations (such as trichloroethylene (TCE)) and EPA's pace in reviewing and potentially revising older standards. Congress last reauthorized SDWA in 1996. Although funding authority for most SDWA programs expired in FY2003, Congress continues to appropriate funds annually for these programs. No broad reauthorization bills have been proposed, as EPA, states, and water systems continue efforts to implement and comply with the requirements of the 1996 law and new regulations.

A long-standing and overarching SDWA issue concerns the cumulative cost and complexity of drinking water standards and the ability of water systems, especially small systems, to comply with standards. The issue of the affordability of drinking water regulations, such as those for arsenic, radium, and disinfection by-products, has merged with the larger debate over what is the appropriate federal role in assisting communities with financing drinking water infrastructure.

Water infrastructure financing legislation has been offered in recent Congresses to authorize increased funding for the DWSRF program and to provide grant assistance for small communities. In the 110th Congress, bills were offered to provide technical, financial, and other compliance assistance to small communities. The Senate Environment and Public Works Committee reported several drinking water bills, including a broad water infrastructure financing bill, S. 3617. None of the legislation was enacted, and the debate continues over the federal role in the funding of projects needed for SDWA compliance, and for water infrastructure improvement in general.

Several bills have been introduced to address the underground injection of carbon dioxide (CO_2) for long-term sequestration as a means of reducing greenhouse gas emissions. P.L. 110-140 (H.R. 6), the Energy Independence and Security Act of 2007, includes carbon sequestration research and development provisions, and specifies that geologic sequestration (GS) activities will be subject to SDWA provision related to protecting underground drinking water sources. In July 2008, EPA proposed regulations under SDWA to provide a national permitting framework for managing the underground injection of CO_2 for commercial-scale GS projects.

Chapter 2 - The events of September 11, 2001, focused heightened attention on the security status of the nation's drinking water supplies and the vulnerability of this critical infrastructure sector to attack. Congress since has enacted security requirements for public water systems and has provided funding for vulnerability assessments, emergency planning, and drinking water research. The Environmental Protection Agency (EPA), the lead federal agency for the water sector, has worked with water utilities, state and local governments, and federal agencies to improve the drinking water security.

The Public Health Security and Bioterrorism Preparedness and Response Act of 2002 (P.L. 107-188) amended the Safe Drinking Water Act to require some 8,400 community water systems to assess vulnerabilities and prepare emergency response plans. It authorized funding for these activities and for emergency grants to states and utilities, and it directed EPA to review methods to prevent, detect, and respond to threats to water safety and infrastructure security. The act did not require water systems to make security upgrades to address potential vulnerabilities. Since FY2002, Congress has appropriated funds annually for EPA to work with states and the water sector to improve the security of drinking water supplies.

In the Homeland Security Act of 2002 (P.L. 107-296), Congress created a Department of Homeland Security (DHS) and gave the DHS responsibility for assessing and protecting the nation's critical infrastructures. However, the act did not transfer EPA's water security functions, and the 2003 Homeland Security Presidential Directive (HSPD-7) affirmed EPA's lead role in protecting the water infrastructure. Under this directive, EPA has responsibility for developing and providing tools and training on improving security to roughly 53,000 community water systems and 16,000 municipal wastewater treatment facilities.

In the 109th Congress, several bills, including a reported bill, S. 2145, proposed to expand water security requirements for certain high-risk water systems. The Department of Homeland Security FY2007 appropriations act (P.L. 109-295) authorized the DHS to regulate for three years high-risk chemical facilities, but the law excluded from coverage drinking water and wastewater treatment facilities.

Although EPA, states, localities, and water utilities have taken steps to address security concerns, the security of the nation's water supplies continues to attract congressional attention. Issues receiving attention have included the status of efforts by the water sector to improve security, whether to increase federal requirements, funding needs for water systems to make security improvements, the relative roles and responsibilities of EPA and DHS regarding the water sector, and the status of research and development of technologies to help water systems detect and address potential biological and chemical contaminants. This chapter reviews governmental and water utility efforts to improve drinking water security.

Chapter 3 - This chapter summarizes the Safe Drinking Water Act (SDWA) and its major programs and regulatory requirements. It excerpts, with several additions, the SDWA chapter of CRS Report RL30798, *Environmental Laws: Summaries of Major Statues Administered by the Environmental Protection Agency (EPA)*, which provides summaries of the principal environmental statutes administered by the Environmental Protection Agency (EPA). This chapter includes the drinking water security provisions added to the SDWA by the Public Heath Security and Bioterrorism Preparedness and Response Act of 2002 (P.L. 107-188).

The Safe Drinking Water Act, Title XIV of the Public Health Service Act, is the key federal law for protecting public water supplies from harmful contaminants. First enacted in 1974 and substantially amended in 1986 and 1996, the act is administered through programs that establish standards and treatment requirements for public water supplies, control underground injection of wastes, finance infrastructure projects, and protect sources of drinking water. The 1974 law established the current federal-state arrangement in which states may be delegated primary implementation and enforcement authority for the drinking water program. The state-administered Public Water Supply Supervision Program remains the basic program for regulating the nation's public water systems, and 49 states have assumed this authority.

The last major reauthorization of the act was done through the Safe Drinking Water Act Amendments of 1996 (P.L. 104-182), which generally authorized appropriations for SDWA programs through FY2003. As with other EPAadministered statutes having expired funding authority, Congress has continued to appropriate funds for the ongoing SDWA programs.

In addition to reviewing key programs and requirements of the SDWA, this chapter includes statistics on the number and types of regulated public water systems. It also provides tables that list all major amendments, with the year of enactment and public law number, and that cross-reference sections of the act with the major U.S. Code sections of the codified statute.

Chapter 4 - The Safe Drinking Water Act (SDWA) Amendments of 1996 authorized a drinking water state revolving loan fund (DWSRF) program to help public water systems finance infrastructure projects needed to comply with federal drinking water regulations and to protect public health. Under the program, states receive capitalization grants to make loans to water systems for drinking water projects and certain other SDWA activities. Since FY1997, Congress has provided more than $10.3 billion for this program, including $829 million for FY2008. As of June 2007, the DWSRF program had provided a total of $12.6 billion in assistance and supported 5,346 projects.

The Environmental Protection Agency's (EPA's) latest (2003) survey of capital improvement needs for public water systems found that water systems need to invest $276.8 billion on infrastructure improvements over 20 years to ensure the provision of safe water. This amount represents a 60% increase over the 1999 survey estimate of $165.5 billion (in 2003 dollars) and reflects funds needed for compliance with several revised regulations (e.g., the arsenic and radium rules), security-related needs, and better reporting of needs for infrastructure rehabilitation and replacement. Key issues include the gap between estimated needs and funding, SDWA compliance costs, and the need for cities to update and maintain water infrastructure, apart from SDWA compliance.

Congress substantially revised the Safe Drinking Water Act with the 1996 SDWA amendments (P.L. 104-182). A key provision, Section 1452, authorized a drinking water state revolving loan fund (DWSRF) program to help public water systems finance improvements needed to comply with federal drinking water regulations and to address the most serious risks to human health. The law authorized EPA to make grants to states to capitalize DWSRFs. States must match 20% of their annual grant and develop intended use plans each year indicating how the allotted funds will be used. States may use the DWSRF to provide loans and other assistance to eligible public water systems for expenditures that EPA has determined will facilitate SDWA compliance or significantly further the act's health protection objectives. Eligible projects include installation and replacement of failing treatment facilities, distribution systems, and certain storage facilities. Projects to replace aging infrastructure are eligible if they are needed to maintain compliance or to further public health protection goals. Projects to consolidate water supplies also may be eligible. This program is patterned after the 1987 Clean Water Act SRF (CWSRF) program for financing municipal wastewater treatment projects.

Public water systems eligible to receive DWSRF assistance include community water systems (whether publicly or privately owned) and not-for-profit noncommunity water systems.[1] States generally may not provide DWSRF assistance to systems that lack the capacity to ensure compliance with the act or that are in significant noncompliance with SDWA requirements, unless these systems meet certain conditions to return to compliance. Systems owned by federal agencies are not eligible. Although the law authorizes assistance to privately owned community water systems, some states have laws or policies that preclude privately owned utilities from receiving DWSRF assistance.

Chapter 5 - Perchlorate is the explosive component of solid rocket fuel, fireworks, road flares, and other products. Used heavily by the Department of Defense (DOD) and related industries, perchlorate also occurs naturally and is present in some organic fertilizer. This soluble, persistent compound has been detected in drinking water supplies, especially in California. It also has been found in milk and many foods. Because of this widespread occurrence, concern over the potential health risks of perchlorate exposure has increased, and some states, water utilities, and Members of Congress have urged the Environmental Protection Agency (EPA) to set a federal drinking water standard for this chemical. Regulatory issues involve the relative health benefits and costs of federal regulation, including environmental cleanup and water treatment costs, both of which are driven by federal and state standards. (California and Massachusetts have set standards.) EPA has spent years assessing perchlorate's health effects and occurrence (including its occurrence in food) to determine whether a national standard is warranted. Interagency disagreements over the risks of perchlorate exposure led several federal agencies to ask the National Research Council (NRC) to evaluate perchlorate's health effects and EPA's risk analyses. In 2005, the NRC issued its report, and EPA adopted the NRC's recommended reference dose (i.e., the expected safe dose) for perchlorate exposure. Subsequent studies raised more concerns about potential effects of low-level exposures, particularly for infants in certain cases. In October 2008, EPA announced its preliminary determination not to regulate perchlorate in drinking water. This chapter reviews perchlorate contamination issues and related developments.

In: Safeguarding the Nation's Drinking Water ISBN: 978-1-60741-234-2
Editors: Juan A. Schrock

Chapter 1

SAFE DRINKING WATER ACT (SDWA): SELECTED REGULATORY AND LEGISLATIVE ISSUES*

Mary Tiemann

SUMMARY

Much progress has been made in assuring the quality of public water supplies since the Safe Drinking Water Act (SDWA) was first enacted in 1974. Public water systems must meet extensive regulations, and public water system management has become a much more complex and professional endeavor. The Environmental Protection Agency (EPA) has regulated some 91 drinking water contaminants, and more regulations are pending. In 2005, EPA reported that the number of systems reporting no violations of drinking water standards reached a new high of 94% in 2003. Despite such progress, however, an array of issues and challenges remain.

In the 110th Congress, key issues have involved infrastructure funding needs, related compliance issues, and concerns caused by detections of unregulated contaminants in drinking water, such as perchlorate and pharmaceuticals and personal care products (PPCPs). Another issue involves the adequacy of existing regulations (such as trichloroethylene (TCE)) and

* This is an edited, reformatted and augmented version of a CRS Report for Congress, Report RL34201, dated November 26, 2008.

EPA's pace in reviewing and potentially revising older standards. Congress last reauthorized SDWA in 1996. Although funding authority for most SDWA programs expired in FY2003, Congress continues to appropriate funds annually for these programs. No broad reauthorization bills have been proposed, as EPA, states, and water systems continue efforts to implement and comply with the requirements of the 1996 law and new regulations.

A long-standing and overarching SDWA issue concerns the cumulative cost and complexity of drinking water standards and the ability of water systems, especially small systems, to comply with standards. The issue of the affordability of drinking water regulations, such as those for arsenic, radium, and disinfection by-products, has merged with the larger debate over what is the appropriate federal role in assisting communities with financing drinking water infrastructure.

Water infrastructure financing legislation has been offered in recent Congresses to authorize increased funding for the DWSRF program and to provide grant assistance for small communities. In the 110th Congress, bills were offered to provide technical, financial, and other compliance assistance to small communities. The Senate Environment and Public Works Committee reported several drinking water bills, including a broad water infrastructure financing bill, S. 3617. None of the legislation was enacted, and the debate continues over the federal role in the funding of projects needed for SDWA compliance, and for water infrastructure improvement in general.

Several bills have been introduced to address the underground injection of carbon dioxide (CO_2) for long-term sequestration as a means of reducing greenhouse gas emissions. P.L. 110-140 (H.R. 6), the Energy Independence and Security Act of 2007, includes carbon sequestration research and development provisions, and specifies that geologic sequestration (GS) activities will be subject to SDWA provision related to protecting underground drinking water sources. In July 2008, EPA proposed regulations under SDWA to provide a national permitting framework for managing the underground injection of CO_2 for commercial-scale GS projects.

INTRODUCTION

The Safe Drinking Water Act[1] (SDWA) is the primary federal law for protecting public water supplies from harmful contaminants. First enacted in 1974, and broadly amended in 1986 and 1996, the SDWA is administered through programs that regulate contaminants in public water supplies, provide funding for infrastructure projects, protect underground sources of drinkingwater, and promote the capacity of water systems to comply with SDWA regulations.

The Environmental Protection Agency (EPA) is the federal agency responsible for administering SDWA; however, the 1974 law established a federal-state structure in which EPA may delegate primary enforcement and implementation authority (primacy) for the drinking water program to states and tribes. The state-administered Public Water Supply Supervision (PWSS) program remains the basic program for regulating public water systems, and EPA has delegated primacy for this program to all states, except Wyoming and the District of Columbia (which SDWA defines as a state); EPA has responsibility for implementing the PWSS program in these two jurisdictions and throughout most of Indian lands.[2]

Since the law was first enacted, much progress has been made in assuring the quality of public water supplies. EPA has regulated some 91 drinking water contaminants, and more regulations are pending. In 2005, EPA reported that the number of public water systems reporting no violations of the health-based standards reached 94% in 2003.[3] However, drinking water safety concerns and challenges remain. EPA and state enforcement data indicate that water systems still incur tens of thousands of violations of SDWA requirements each year. Although these violations primarily involve monitoring and reporting requirements, they also include thousands of violations of standards and treatment techniques. Moreover, monitoring and reporting violations create uncertainty as to whether systems actually met the applicable health-based standards. While noting increased compliance levels in the 2005 report, EPA estimated that only 65% of violations of health-based standards and 23% of violations of monitoring and reporting requirements were reported to the EPA database, thus increasing uncertainty as to the quality of water provided by many systems. EPA and the states have resolved some data quality and reporting problems, and efforts continue to address this issue. In 2008, EPA reported that 91.5% of the total population served by community water systems received drinking water that met all health-based standards during FY2007, up from 89.4% in FY2006.[4] Concern also exists over the potential health effects of drinking water contaminants for which standards have not been set, such as perchlorate and methyl tertiary butyl ether (MTBE). The act requires EPA to continually evaluate contaminants that may be candidates for regulation; however, EPA's perceived lack of action on contaminants of concern has generated criticism in Congress and elsewhere.

Last Major Reauthorization and Amendments

Congress last broadly revised the act with the Safe Drinking Water Act Amendments of 1996 (P.L. 104-182). These changes resulted from a multi-year effort to amend a statute that was widely criticized as having too little flexibility, too many unfunded mandates, and an arduous but unfocused regulatory schedule. Among the key provisions, the 1996 amendments authorized a drinking water state revolving loan fund (DWSRF) program to help public water systems finance projects needed to comply with SDWA regulations. The amendments also established a process for selectingcontaminants for regulation based on health risk and occurrence, gave EPA some added flexibility to consider costs and benefits in setting most new standards, and established schedules for regulating certain contaminants (such as *Cryptosporidium*, disinfection byproducts, arsenic, and radon).

The 1996 law added several provisions aimed at building the capacity of water systems (especially small systems) to comply with SDWA regulations. It imposed many new requirements on the states. For example, the amendments require state programs for source water assessment, operator certification and training, and compliance capacity development. The law also requires community water systems to provide customers with annual "consumer confidence reports" that contain information on regulated contaminants found in the local drinking water. Congress authorized appropriations for most SDWA programs through FY2003, and although most funding authorities have expired, broad reauthorization bills have not been proposed, as EPA, states, and public water systems remain focused on meeting the requirements of the 1996 amendments.

In 2002, drinking water security provisions were added to the SDWA through the Public Health Security and Bioterrorism Preparedness and Response Act of 2002 (P.L. 107-188). Title IV of that act included requirements for community water systems serving more than 3,300 people to conduct vulnerability assessments and prepare emergency response plans. The law also required the EPA to conduct research on preventing and responding to terrorist or other attacks.[5]

Regulated Public Water Systems

Federal drinking water regulations apply to some158,200 privately and publicly owned water systems that provide piped water for human consumption to at least 15 service connections or that regularly serve at least 25 people. (The law does not apply to private residential wells.) Of these systems, 52,837 are *community water systems* (CWSs) that serve most people in the United States — a total residential population of roughly 282 million year-round. All SDWA regulations apply to these systems. Nearly 19,200 systems are *non-transient, non-community water systems* (NTNCWSs), such as schools or factories, that have their own water supply and serve the same people for more than six months but not year-round. Most drinking water requirements apply to these systems. Another 86,210 systems are *transient non-community water systems* (TNCWSs) (e.g., campgrounds and gas stations) that provide their own water to transitory customers. TNCWSs generally are required to comply only with regulations for contaminants that pose immediate health risks (such as microbial contaminants), with the proviso that systems that use surface water sources must also comply with filtration and disinfection regulations.

Of the nearly 53,000 community water systems, roughly 83% serve 3,300 or fewer people. While large in number, these systems provide water to just 9% of the population served by all community systems. In contrast, 8% of community water systems serve more than 10,000 people, and they provide water to 81% of the population served. Fully 85% (16,348) of non-transient, non-community water systems and 97% (83,351) of transient noncommunity water systems serve 500 or fewer people. These statistics give some insight into the scope of financial, technological, and managerial challenges many public water systems face in meeting a growing number of complex federal drinking water regulations. **Table 1** provides statistics for community water systems.

SAFE DRINKING WATER ACT ISSUES

Current drinking water quality issues include the gap between infrastructure funding needs and spending; the capacity of public water systems, especially small systems, to comply with a growing set of complex standards; and contamination of water supplies by unregulated contaminants, such as perchlorate and methyl tertiary butyl ether (MTBE). An emerging issue concerns proposals for large-scale storage of carbon dioxide deep underground to mitigate greenhouse gas emissions and the potential impacts this activitymight haveon groundwater quality. In July 2008, EPA proposed regulations to manage geologic sequestration under the authority of the Safe Drinking Water Act's groundwater protection provisions.

In the 110th Congress, bills were offered to address drinking water issues, such as small system infrastructure funding needs (S. 1933 and S. 199) and compliance problems facing small communities (S. 1429 and H.R. 2141). Various bills also proposed to address contamination of water supplies by specific contaminants, including perchlorate (S. 24, S. 150, and H.R. 1747), lead (H.R. 2076), and arsenic and other naturally occurring contaminants (H.R. 2141). The Senate Environment and Public Works Committee held hearings on selected EPA activities, including perchlorate regulation. The House Natural Resources Committee held a hearing on the impacts of perchlorate contamination on groundwater supplies. The House Energy and Commerce Committee held a hearing on the contamination of drinking water at Camp Lejeune by trichloroethylene (TCE), related health and cleanup issues, and EPA's pace in updating its 1989 drinking water standard for TCE. Relatedly, S. 1911 and H.R. 5527 would have required EPA to promptly issue a new TCE standard. The Senate Environment and Public Works Committee reported perchlorate bills, S. 24 (S.Rept. 110-483) and S. 150 (S.Rept. 110-484), TCE legislation (S. 1911, S.Rept. 110-486), and legislation to authorize a small system grant program (S. 1933, S.Rept. 110-475). The committee also reported S. 3617 (S.Rept. 110-509), the Water Infrastructure Financing Act, which would have authorized a grant program and increased funding for drinking water and wastewater infrastructure. Other reported bills included S. 2970 (S.Rept. 110-487) to help water utilities develop and implement climate change adaptation policies, and S. 199 (S.Rept. 110-476), to increase the authorization of appropriations for water and wastewater grants for Alaska's rural and Native villages. None of the bills was enacted.

In the 1996 SDWA amendments, Congress authorized appropriations for most SDWA programs through FY2003; however, broad SDWA reauthorization efforts were not on the agenda in the 110th Congress. As with other EPA-administered statutes having expired funding authority, Congress has continued to appropriate funds annually for SDWA programs.

Regulating Drinking Water Contaminants

Contaminant Candidate List

The SDWA Amendments of 1996 directed EPA to publish, every five years, a list of unregulated contaminants that are known or anticipated to occur in public water systems and that may require regulation. EPA published contaminant candidate lists (CCLs) in 1998 (CCL 1) and in 2003 (CCL 2). In February 2008, EPA published for public comment a draft CCL 3 that contains 93 chemicals or chemical groups and 11 microbiological contaminants (73 *Fed. Reg.* 9627). The list includes commercial and agricultural chemicals, biological toxins, disinfection byproducts, and pathogens; 16 chemicals, including perchlorate, were carried over from CCL 2. EPA screened some 7,500 chemicals and microbes and selected 104 candidates for the draft CCL 3. However, as discussed below, the list does not include any pharmaceuticals, and EPA is now reviewing the adequacy of its screening process, which was recommended by the National Academy of Sciences.

Regulatory Determinations

Starting in 2001, and every five years thereafter, the EPA is required to determine whether or not to regulate at least five of the listed contaminants. The act requires the EPA to evaluate contaminants that present the greatest health concern, and then to regulate those contaminants that occur at concentration levels and frequencies of public health concern, where regulation presents a meaningful opportunity for health risk reduction.

Table 1. Size Categories of Community Water Systems

System size (population served)	Number of community water systems	Population served (millions)	Percentage of community water systems	Percentage of population served
Very small (25-500)	29,666	4.93	56%	2%
Small (501-3,300)	14,389	20.85	27%	7%
Medium (3,301-10,000)	4,748	27.51	9%	10%
Large (10,001-100,000)	3,648	102.75	7%	36%
Very large (>100,000)	386	126.30	1%	45%
Total	52,837	282.3	100%	100%

Source: Adapted from US Environmental ProtectionAgency, *Factoids: Drinking Water and Ground Water Statistics for 2005.*

In July 2008, EPA published final regulatory determinations for 11 contaminants from the CCL 2. All of the determinations were decisions not to regulate. In making these determinations, EPA noted that the data indicated that the contaminants either did not appear to occur in public water systems, or had low levels of occurrence at levels of healthconcern, and that regulating the contaminants did not present a meaningful opportunity for health risk reduction. For those contaminants with low occurrence frequencies, EPA is updating the health advisories to reflect new information or to include information on degradation byproducts.[6]

The agency did not make determinations for two chemicals that have been detected in numerous water supplies and have received considerable congressional attention: perchlorate and MTBE. EPA noted that a regulatory determination will soon be published for perchlorate, and that a decision was not made for MTBE because the health risk assessment for MTBE is being revised.

Unregulated Contaminant Monitoring

The 1996 amendments directed EPA to establish criteria for a program to monitor unregulated contaminants. This monitoring program enables EPA to collect data for contaminants that are not regulated but are suspected to be present in drinking water. Every five years, EPA is required to identifyasmanyas30 contaminants to be monitored. This list is largely based on the Contaminant Candidate List. All systems serving more than 10,000 people and a sample of smaller systems must monitor for the contaminants. The resulting data are added to the National Contaminant OccurrenceDatabase(NCOD). EPA published the first unregulated contaminant monitoring rule (UCMR 1) in 1999 requiring monitoring for 26 chemicals. In January 2007, EPA issued the second rule (UCMR 2), requiring systems to monitor for 25 chemicals over a 12-month period between 2008 through 2010.[7] EPA had included perchlorate on the draft UCMR 2 list, but deleted it from the final list. EPA stated that it has sufficient perchlorate occurrence data, but some advocates of perchlorate regulation were critical of EPA's decision not to require further monitoring.

Standard-Setting

As amended in 1996, the act directs EPA to promulgate a National Primary Drinking Water Regulation for a contaminant if the Administrator determines that the following three criteria are met:

- the contaminant may have adverse health effects;
- it is known, or there is a substantial likelihood, that the contaminant will occur in public water systems with a frequency and at levels of public health concern; and
- its regulation presents a meaningful opportunity for health risk reduction for persons served by public water systems.

Drinking water regulations generally include numerical standards that establish the highest level of a contaminant that may be present in water supplied by public water systems. Where it is not economically or technically feasible to measure a contaminant at very low concentrations, EPA may establish a treatment technique in lieu of a standard.

Developing a drinking water regulation is a complex process, and EPA must address technical, scientific, and economic issues. The agency must (1) estimate the extent of occurrence of a contaminant in sources of drinking water nationwide; (2) evaluate the potential human exposure and risks of adverse health effects to the general population and to sensitive subpopulations; (3) ensure that analytical methods are available for water systems to use in monitoring for a contaminant; (4) evaluate the availability and costs of treatment techniques that can be used to remove a contaminant; and (5) assess the impacts of a regulation on public water systems, the economy, and public health. Regulation development typically is a multi-year process. EPA may expedite procedures and issue interim standards to respond to urgent threats to public health.

After reviewing health effects studies, EPA sets a nonenforceable maximum contaminant level goal (MCLG) at a level at which no known or anticipated adverse health effects occur and that allows an adequate margin of safety. EPA also considers the risk to sensitive subpopulations, such as infants and children. For carcinogens and microbes, EPA generally sets the MCLG at zero. Because MCLGs are based only on health effects and not on analytical detection limits or the availability or cost of treatment technologies, they may be set at levels that are not technically feasible for water systems to meet.

Once the MCLG is established, EPA then sets an enforceable standard, the maximum contaminant level (MCL). The MCL generally must be set as close to the MCLG as is "feasible" using the best technology or other means available, taking costs into consideration (SDWA §1412(b)).[8] The act does not discuss how EPA should consider cost in determining feasibility; consequently, EPA has relied on legislative history for guidance. Congress last addressed this issue in the Senate report accompanying the 1996 amendments, which stated that "feasible" means the level that can be reached by large, regional drinking water systems applying best available treatment technology. The Senate committee report explained that this approach is used because 80% of the population receives its drinking water from large community water systems, and thus, safe water can be provided to most of the population at very affordable costs.[9]

However, because standards are based on cost considerations for large systems, Congress expected that standards could be less affordable for smaller systems. In 1996, Congress expanded the act's variance and exemption provisions to give small systems some added compliance flexibility. (See the discussion below on Small System Issues.) Congress further revised the act to require EPA, when proposing a standard, to publish a determination as to whether or not the benefits of a proposed standard justify the costs. If EPA determines that the benefits do not justify the costs, EPA, in certain cases, may promulgate a standard that is less stringent than the feasible level and that "maximizes health risk reduction benefits at a cost that is justified by the benefits."[10] EPA used this authority to establish new standards for arsenic and radium.

Recent and Pending Rules

EPA's rulemaking activities include a January 2006 rule package that expanded existing requirements to control pathogens (especially *Cryptosporidium*) and disinfectants (e.g., chlorine) and their byproducts (e.g., chloroform). These rules, the Long Term 2 Enhanced Surface Water Treatment Rule (LT2 Rule) and the Stage 2 Disinfectant and Disinfection Byproduct Rule (Stage 2 DBP), complete a series of statutorily mandated rules that impose increasingly strict controls on the presences of pathogens and disinfectants and their byproducts in water systems.[11] EPA promulgated a related Ground Water Rule to establish disinfection requirements for systems relying on ground water. In the past several years, EPA also issued standards for several radionuclides, including uranium and radium, and a revised standard for arsenic. These rules are expected to reduce an array of health risks for consumers, but they have potentially significant costs for the communities that must expand treatment facilities to comply with the standards.

In September 2007, EPA completed targeted revisions to the Lead and Copper Rule (LCR). The revisions are intended to address weaknesses identified during a nationwide review of the rule, following the discovery of high lead levels in Washington, DC, tap water in 2004.[12] The changes involved regulatory requirements for monitoring, treatment, customer notification, and lead service line replacement. More comprehensive revisions to the LCR are under way. Among ongoing rulemakings, EPA has been working to finalize a radon rule (proposed in 1999), and has been evaluating numerous contaminants, including perchlorate and MTBE, for possible regulation. **Table 2** reviews the purposes and status of several recently completed or proposed drinking water regulations and guidelines.

Table 2. Recent and Pending Regulatory Actions

Regulatory Action	Date Published (*Fed. Reg.* Notice)	Purpose
Revisions to Lead and Copper Rule (LCR)	10/10/2007 (72 *Fed. Reg.* 57781) Final	EPA promulgated targeted changes to the LCR to improve implementation in the areas of monitoring, treatment, customer awareness, and lead service line replacement, to better control exposures to lead in drinking water. The revisions do not affect the lead MCLG or action level, or the rule's basic requirements. (A comprehensive revision of the rule is under way.)
Unregulated Contaminant Monitoring Rule (UCMR 2)	1/4/2007 (72 *Fed. Reg.* 367) Final	SDWA requires EPA to publish every five years a list of unregulated contaminants to be monitored. This second UCMR requires monitoring of 25 chemi-cals during 2008-2010. These data provi-de the main occurrence and exposure data for EPA to determine whether to regulate the contaminants. (Perchlorate was included in the first UCMR and in the draft, but not final, UCMR 2.)
Ground Water Rule (GWR)	11/8/2006 (71 *Fed. Reg.* 65574) Final	The 1996 amendments directed EPA to require· disinfection for all public water systems, including all surface water systems and, as necessary, ground water systems to provide greater protection against microbial pathogens.
Proposed Revision of National Affordability Methodology and Methodology to Identify Variance Technologies	3/2/2006 (71 *Fed. Reg.* 65573)	EPA proposed options for revising its criteria for determining whether a technology needed to comply with a standard is affordable for small systems and for revising its methodology for determining if an affordable variance technology protects public health. As provided for in the 1996 amendments, states may grant variances to small systems for standards that EPA determines are unaffordable. Under the current criteria, no small system variances are available.

Table 2. (Continued)

Regulatory Action	Date Published (*Fed. Reg.* Notice)	Purpose
Long-Term 2 Enhanced Surface Water Treatment Rule (LT2 Rule)	1/5/2006 (71 *Fed. Reg.* 653) Final	Supplements existing rules by increasing *Cryptosporidium* treatment requirements for higher risk systems. Contains provisions to reduce risks from uncovered finished water reservoirs and to ensure that systems maintain microbial protection when they act to decrease the formation of disinfection byproducts (DBPs).
Stage 2 Disinfectants and Disinfection By-Products Rule (DBPR)	1/4/2006 (71 *Fed. Reg.* 387) Final	Builds on existing rules to strengthen requirements for higher risk systems to reduce potential health risks from DBPs in drinking water, which form when disinfectants are used to control microbial pathogens. Tightens monitoring require-ments for 2 groups of DBPs, trihalomethanes (TTHM) and haloacetic acids (HAA5). (This rule was issued with the LT2 Rule to address concerns about risk tradeoffs between pathogens and disinfection byproducts.)
Proposed Radon Rule	11/2/1999 (64 *Fed. Reg.* 59245)	As provided for in P.L. 104-182, EPA proposed a multimedia approach to reducing radon risks in indoor air (the biggest exposure source) while protecting public health from radon in drinking water.[a] EPA proposed an alternative standard (AMCL) and requirements for multimedia mitigation (MMM) programs to address radon in indoor air. A community water system (CWS) may comply with the AMCL if the state develops an MMM program or the CWS develops a state-approved MMM program. EPA also proposed a stricter radon MCL to apply in states that do not implement MMM programs.

a. Most of the risk from exposure to radon in drinking water comes from breathing radon released from water (e.g., during showering or cooking) and not from ingesting the water. EPA estimates that 1-2% of the radon in indoor air comes from drinking water. Most of the radon present in indoor air seeps into homes and other buildings from underlying soil.

Perchlorate in Drinking Water

Perchlorate is the explosive component of solid rocket fuel, fireworks, road flares, and other products. Used heavily by the Department of Defense (DOD) and related industries, perchlorate also occurs naturally (including in areas of the Southwest) and is present in some organic fertilizer. This compound has been detected in sources of drinking water for more than 11 million people, usually at low levels. It also has been found in milk, fruits, vegetables, and bread. Perchlorate is known to disrupt the uptake of iodine in the thyroid, potentially affecting thyroid function. A key concern is that, if sufficiently severe, impaired thyroid function in pregnant or nursing women may impair brain development in fetuses and nursing infants. Because of this concern, bills have been repeatedly introduced in recent Congresses to require EPA to set a drinking water standard for perchlorate.[13]

Over the past decade, EPA has been evaluating perchlorate to determine whether a federal drinking water standard is warranted. Regulatory issues have involved the relative risk reduction benefits and costs of federal regulation, including environmental cleanup and water treatment costs, both of which are driven by federal and state standards. Another question has concerned the amount of perchlorate exposure that is attributable to food compared to drinkingwater. Under SDWA, EPA must regulate a contaminant if the Administrator determines that the contaminant occurs at a frequency and level of public health concern, and that its regulation presents a meaningful opportunity for reducing health risks. Uncertainty about the health risks associated with exposure to perchlorate at low levels slowed EPA efforts to determine whether to establish a standard and, relatedly, environmental cleanup standards for use at Superfund and other contaminated sites.

In the absence of a federal standard, states have begun to adopt their own measures. Massachusetts established a drinking water standard for perchlorate of 2 ppb in 2006, and California adopted a standard of 6 ppb. Several states have issued health goals or advisory levels, ranging from 1 ppb in Maryland (advisory level) and New Mexico (drinking water screening level) to 4 ppb in Texas (residential groundwater cleanup level).

EPA identified perchlorate as a candidate for regulation in 1998, but concluded that information was insufficient at that time to make a regulatory determination. The agency listed perchlorate as a priority for further research on health effects and treatment technologies and for collecting occurrence data. In 1999, EPA required water systems to monitor for perchlorate under the Unregulated Contaminant Monitoring Rule (UCMR) to determine the frequency and levels at which it is present in public water supplies nationwide. In monitoring conducted under the UCMR, perchlorate was detected in 153 public water systems in 25 states, out of 3,600 water systems sampled nationwide. In August 2005, EPA proposed a second UCMR that included perchlorate for additional monitoring between 2007 and 2011. Most commentors did not support another round of perchlorate monitoring. Many felt that further monitoring would impose costs but would not likely yield much beneficial information. In the final rule, EPA announced that it had collected sufficient occurrence data for perchlorate and that further monitoring was not needed (72 *Federal Register* 367, January 4, 2007).

The health effects assessments surrounding EPA's efforts to regulate perchlorate have been controversial. In 2002, EPA issued a draft risk assessment that concluded that potential human health risks of perchlorate exposure include effects on the developing nervous systems and thyroid tumors. The findings were based on rat studies that observed tumors and adverse effects in fetal brain development. This controversial draft assessment included a revised draft reference dose (RfD) intended to protect the most sensitive groups against these effects. That dose roughly translated to a drinking water standard of 1 part per billion (ppb). EPA's 1999 draft level had translated to a standard of roughly 32 ppb.

Because the reference dose provides the basis for determining the level at which a drinking water standard is set, and because these standards are, in turn, the basis of environmental cleanup standards, DOD and other perchlorate users and manufacturers have followed EPA's perchlorate risk assessment efforts closely. As a result of interagency debate over the draft assessment, in 2003, EPA, the DOD, NASA, the Office of Management and Budget, and other federal agencies asked the National Research Council (NRC) to review the science for perchlorate and EPA's draft risk assessment.

The NRC released its study in January 2005.[14] The NRC committee broadly agreed with several of EPA's findings, but suggested several changes to the draft risk assessment. Among other findings, the committee concluded that rat studies are of limited use for assessing human health risk associated with perchlorate exposure, and the committee recommended that EPA base its assessment on human data. The NRC calculated an RfD for perchlorate that incorporated an uncertainty factor intended to protect the most sensitive populations. EPA adopted the NRC's recommended RfD, which translates to a drinking water equivalent level of 24.5 ppb. (If EPA were to develop an MCL, it would adjust this number to take into account the amount of perchlorate exposure that comes from other sources, especially food.)

Despite the NRC recommendations, substantial disagreement has persisted regarding what level of exposure is safe, especially for fetuses and infants, and what drinking water standard is appropriate. Several studies have indicated that thyroid changes occur in humans at significantly higher concentrations of perchlorate than the amounts typically observed in water supplies.[15] However, a 2006 study by the Centers for Disease Control and Prevention (CDC) of a representative sample of the U.S. population found that environmental exposures to perchlorate have an effect on thyroid hormone levels in women with iodine deficiency. The researchers reported that more than one-third of the 1,111 women in the study were iodine deficient, and the median level of urinary perchlorate measured in the women was 2.9 ppb.[16]

On October 3, 2008, EPA announced a preliminary determination not to set a standard for perchlorate, noting that less than 1% of water systems have perchlorate levels above EPA's health reference level (discussed below). EPA concluded that perchlorate failed to meet two of SDWA's regulatory criteria (i.e., that a contaminant occurs frequently at levels of health concern, and that establishing a national drinking water standard would provide a "meaningful opportunity for health risk reduction"). In response, EPA's Science Advisory Board's (SAB's) Drinking Water Committee argued that, given perchlorate's occurrence and well-documented toxicity, EPA must have a compelling basis to support a determination not to regulate. The SAB requested more time to review the new model used by EPA, and to comment on the preliminary determination. If the final determination is not to regulate, EPA will develop a nonenforceable health advisory to guide state and other officials on health effects, monitoring and treatment technologies, and expected safe exposure levels.

In the 110th Congress, perchlorate again was on the agenda. S. 24 was introduced following EPA's decision not to require further monitoring for perchlorate as an unregulated contaminant (as discussed above under "Regulatory Determinations"). This bill would have required water systems to test for perchlorate and disclose its presence in annual consumer confidence reports. Similarly, in the absence of a decision by EPA to regulate perchlorate, S. 150 proposed to require EPA to issue a perchlorate health advisory within 90 days of enactment, and to promptly establish a drinking water standard. In September 2008, the Senate Environment and Public Works Committee reported S. 24 (S.Rept. 110-483) and S. 150 (S.Rept. 110-484). In the House, H.R. 1747 would have required EPA to issue a perchlorate standard. In 2007, the Environment and Hazardous Materials Subcommittee of the House Committee on Energy and Commerce held markup and forwarded H.R. 1747 to the full committee. No further action occurred on these bills.

Pharmaceuticals in Drinking Water

As monitoring technologies have become available and testing has increased, traces of more pharmaceuticals and personal care products (PPCPs) have been detected in surface waters and drinking water supplies. Pharmaceuticals include prescription drugs, veterinary drugs, and over-the-counter medicines. Personal care products cover a broad spectrum and include cosmetics, hair products, sun-screens, fragrances, anti-bacterial soaps, and vitamins. These chemicals are released to the environment in various ways, including elimination of human and animal waste, disposal of unused medicines down the toilet, veterinary drug usage, hospital waste disposal, and industrial discharges.

Although significant research is being conducted, much is unknown about the occurrence and movement of PPCPs in the environment, their occurrence in drinking water supplies, or about the potential health risks from exposure to PPCPs at extremely low levels through drinking water. Nonetheless, the detection of pharmaceuticals and related products in public water supplies generates concern, because many of these products are specifically designed to have a biological effect in humans, animals, and/or plants. Pharmaceuticals often contain chemical compounds that can affect the endocrine system by altering, mimicking, or impeding the function of hormones. Such endocrine disrupting chemicals (EDCs) have the potential to affect growth, development, reproduction, and metabolism. Over the past decade, scientists and regulators have become increasingly concerned about the effects that exposures to low levels of PPCPs may be having on aquatic organisms, and also potentially on human health.[17]

The U.S. Geological Survey (USGS) and EPA have identified a wide array of research needs and gaps that, if addressed, would help delineate the scope of environmental and human health issues that mightresultfromthe presence of PPCPs in the environment. The USGS has conducted research on the occurrence of hormones, pharmaceuticals, and other wastes in residential, industrial, and agricultural wastewater, and has found that a broad range of these chemicals occur commonly downstream from large urban areas and concentrated animal production areas.[18]

EPA has been conducting and supporting numerous PPCP research projects in several areas, including the relative importance of different sources of PPCPs in the environment (e.g., veterinary vs. human medicine), how PPCPs move through the environment, human exposure pathways, ecological exposure pathways, monitoring and detection tools, assessment of potential human health effects, and assessment of potential ecological effects. Research is also being conducted to evaluate the ability of drinking water treatment technologies to remove various PPCPs.

Ecological research has received particular attention because exposure risks for aquatic life have been considered to be much greater than those for humans.[19] Nonetheless, a key research issue concerns the possible health risks from exposure to very low doses of the myriad chemicals found in PPCPs. Because PPCPs occur in the environment at low concentrations, their effects may be subtle. Among other research gaps, EPA has identified a need to develop tests that can detect more subtle health effects.

The agency is also conducting a study to determine the amount of PPCPs that are discharged to wastewater treatment plants from various sources. As part of this study, EPA is evaluating how hospitals and other institutions dispose of unused medications.[20] Other research projects address the development of analytical methods to determine the source and fate of PPCPs in the environment.

As noted above, EPA proposed its third list of unregulated contaminants being considered for regulation in February 2008. This Contaminant Candidate List 3 (CCL 3) contains 104 contaminants, none of which are pharmaceuticals. Following recent reports of the detection of pharmaceuticals and commonly used over-the-counter drugs in the drinking water supplies of 24 large community water systems, EPA has asked its Science Advisory Board (SAB) and stakeholders to evaluate and comment on the contaminant candidate screening and selection process to determine whether the process requires revision.[21]

Because of ecological concerns, as well as human health concerns, regulating contaminants in drinking water represents only part of the response to this multi-faceted problem. Recognizing that people and animals will continue to take and use pharmaceutical products, water suppliers and other stakeholders consider changes at wastewater treatment plants to be a key part of the solution.

The Association of Metropolitan Water Agencies (AMWA), which represents the largest publicly owned water systems, has made several recommendations to address this emerging drinking water issue. Among these recommendations, the AMWAstrongly encouraged EPA to make research on treatment technologies a high priority, and urged water utilities to inform consumers of efforts to monitor and remove pharmaceuticals from water sources. AMWA also called for EPA and the Food and Drug Administration (FDA) to determine whether the presence of trace amounts of pharmaceuticals results in short-term or long-term effects on health and the environment, recommended that the federal government take the lead in developing a national program for disposing of unused prescriptions, and called for animal feeding operations to reduce their contributions of antibiotics and steroids into water supplies.[22]

H.R. 6820, introduced on August 1, 2008, proposed to require EPA to work with other agencies to conduct a study on the presence of PPCPs in the nation's drinking water supplies. The bill would have required the Administrator to submit reports to Congress on the types, levels, and sources of PPCPs found in drinking water; the human health and ecological effects of PPCPs; monitoring and removal technologies; disposal methods; and other information.

Drinking Water Infrastructure Needs and Funding

A persistent SDWA issue concerns the ability of water systems to construct or upgrade infrastructure to comply with drinking water regulations and, more broadly, to ensure the provision of a safe and reliable water supply. In the 1996 amendments, Congress responded to growing complaints about the act's unfunded mandates and authorized a drinking water state revolving loan fund (DWSRF) program to help water systems finance infrastructure projects needed to meet drinking water standards and address the most serious health risks. The program authorizes EPA to award annual capitalization grants to states. States then use their grants (plus a 20% state match) to provide loans and other assistance to public water systems. Communities repay loans into the fund, thus replenishing the fund and making resources available for projects in other communities. Eligible projects include installation and replacement of treatment facilities, distribution systems, and some storage facilities. Projects to replace aging infrastructure are eligible if they are needed to maintain compliance or to further public health protection goals.[23]

The SDWA authorized appropriations for the DWSRF program totaling $9.6 billion, including $1 billion for each of FY1995 through FY2003. Since FY1997, Congress has appropriated more than $10.3 billion for this program, including $829.0 million (after applying the 1.56% across-the-board rescission) for FY2008. The President requested $842.2 million for FY2009. Specific EPA appropriations were not passed, and the consolidated appropriations act for FY2009 (P.L. 110-329) generally extended funding for

EPA programs at FY2008 levels through March 6, 2009. (**Table 3** lists funding levels for the DWSRF program since its inception.)

Through June 2007, the EPA had awarded $8.13 billion in capitalization grants, which, when combined with the state match, bond proceeds, loan principal repayments, and other funds, amounted to $13.9 billion in DWSRF funds available for loans and other assistance. Also, as of June 2007, 5,346 projects had received assistance, and total assistance provided by the program reached $12.63 billion.

The DWSRF program is well-regarded, but many state and local officials and interest groups have argued that greater investment in water infrastructure is needed. EPA's 2003 drinking water infrastructure needs survey found that systems need to invest $276.8 billion in infrastructure improvements over 20 years to comply with drinking water regulations and to ensure the provision of safe water.[24] The survey includes funds needed for compliance with several recent rules (including the arsenic rule and the disinfectants and disinfection byproducts rules) and several proposed rules (including radon). The survey also identified $1 billion in security-related needs. All infrastructure projects in the needs assessment promote the health objectives of the act, but only $45.1 billion (16.3%) of the total need is attributable to SDWA compliance. Although aging, deteriorated infrastructure often poses a threat to drinking water safety, these needs occur independently of federal mandates.

EPA also has prepared a broader municipal wastewater and drinking water infrastructure funding gap analysis, which identified potential funding gaps between projected needs and spending from 2000 through 2019.[25] This analysis estimated the potential 20-year funding gap for drinking water and wastewater infrastructure capital and operations and maintenance (O&M), based on two scenarios: a "no revenue growth" scenario and a "revenue growth" scenario that assumed infrastructure spending would increase 3% per year. Under the "no revenue growth" scenario, EPA projected a funding gap for drinking water capital investment of $102 billion ($5 billion per year) and an O&M funding gap of $161 billion ($8 billion per year).

Using revenue growth assumptions, EPA estimated a 20-year capital funding gap of $45 billion ($2 billion per year), and no gap for O&M. In response to the Gap Analysis, EPA's FY2004 budget request proposed that funding for the DWSRF program be continued at a level of $850 million annually through FY2018. EPA explained that this funding level would allow DWSRFs to revolve at a cumulative level of $1.2 billion (more than double the previous goal of $500 million) and would help close the funding gap for drinking water infrastructure needs.

Other assessments also have found a funding gap. In 2000, the Water Infrastructure Network (WIN) (a coalition of state and local officials, water providers, environmental groups and others) reported that over the next 20 years, water and wastewater systems need to invest $23 billion annually more than current investments to meet SDWA and Clean Water Act health and environmental priorities and to replace aging infrastructure. WIN and other groups have proposed multibillion dollar investment programs for water infrastructure. Others, however, have called for more financial self-reliance within the water sector.

Table 3. Drinking Water State Revolving Fund Program Funding, FY1997-FY2009 (in millions of dollars, nominal dollars and adjusted for inflation)

Fiscal Year	Authorizations	Appropriations	Adjusted for Inflation in 2007 Dollars
1997	$1,000.0	$1,275.0	$1,594.6
1998	$1,000.0	$725.0	$895.9
1999	$1,000.0	$775.0	$945.2
2000	$1,000.0	$816.9	$976.6
2001	$1,000.0	$823.2	$961.4
2002	$1,000.0	$850.0	$974.1
2003	$1,000.0	$844.5	$948.6
2004	—	$845.0	$925.1
2005	—	$843.2	$894.5
2006	—	$837.5	$859.9
2007	—	$837.5	$837.5
2008	—	$829.0	est. $813.3
2009 (req.)	—	$842.2	est. $809.9

Sources: Prepared by CRS using information from the following sources: FY1997-FY2000 and FY2002 enacted amounts are from the enacted appropriations bills for those fiscal years. FY2001enacted amount is the prior year enacted amount specified in EPA's FY2002 congressional budget justification. FY2003-FY2004 enacted amounts are from EPA's Office of Water. FY2005-FY2006 enacted amounts are prior year enacted amounts specified in House Appropriations Committee reports on subsequent year appropriations bills. FY2007 and FY2008 enacted amounts, and FY2009 President's request, are as reported to CRS by the House Appropriations Committee. All enacted amounts reflect rescissions. Nominal dollar amounts were converted into 2007 dollar values usingthe GDP Chained Price Index from the Office of Managementand Budget, *Budget of the United States Government for Fiscal Year 2009*, Historical Tables.

In the 109th Congress, the Senate Environment and Public Works Committee reported the Water Infrastructure Financing Act, S. 1400 (S.Rept. 109-186), which would have amended the SDWA and the Clean Water Act to reauthorize both SRF programs (authorizing $15 billion over five years for the DWSRF). The bill also would have directed EPA to establish grant programs for small or economically disadvantaged communities for critical drinking water and water quality projects; authorized loans to small systems for preconstruction, short-term, and small project costs; and directed EPA to establish a demonstration program to promote new technologies and approaches to water quality and water supply management. At markup, the committee adopted an amendment to apply Davis-Bacon prevailing wage requirements, in perpetuity, to projects receiving DWSRF assistance. The Davis-Bacon measure remained contentious, and S. 1400 received no further action.

The 110th Congress returned to water infrastructure funding and needs issues. On September 26, 2008, the House passed an emergency supplemental appropriations bill, H.R. 7110 (the Job Creation and Unemployment Relief Act of 2008). This bill proposed to provide $1 billion for the DWSRF program, and another $6.5 billion for the Clean Water SRF program. Also in September, the Senate Environment and Public Works Committee reported several bills that would have authorized funding for drinking water infrastructure. S. 3617 (S.Rept. 110-509), the Water InfrastructureFinancing Act, which was similar to the committee bill from the 109th Congress, would have authorized more funding for drinking water and wastewater SRF programs (authorizing $15 billion over five years for the DWSRF), and would have created a grant program at EPA for small or economically disadvantaged communities for critical drinking water and water quality projects. S. 3617 included a Davis-Bacon prevailing wage provision, requiring that prevailing wage requirements would apply to all projects financed in whole or part through an SRF.) The other reported bills included S. 1933 (S.Rept. 110-475), which would have created a grant program for small drinking water systems and would have authorized $750 million annually for FY2008 through FY2014; and S. 199 (S.Rept. 110-476), to increase the authorization of appropriations for water and wastewater grants for Alaska's rural and Native villages. None of the bills was enacted.

In the face of uncertainty over increased federal assistance for water infrastructure, EPA, states, communities, and utilities have been examining alternativemanagement and financing strategies to address SDWA compliance costs and broader infrastructure maintenance and repair costs. Such strategies include establishing public-private partnerships (privatization options range from contracting for services to selling system assets), improving asset management, and adopting full-cost pricing for water services. Still, these strategies may be of limited use to many small and/or economically disadvantaged communities, and stakeholders are likely to continue to urge Congress to increase funding for water infrastructure.[26]

Small Systems Issues

An issue that has received considerable attention concerns the financial, technical, and managerial capacity of small systems to comply with SDWA regulations. Roughly 84% (44,000) of the nation's 52,800 community water systems are small, serving 3,300 persons or fewer, and 57% (30,000) of the community water systems serve 500 persons or fewer. Many small systems face challenges in complying with SDWA rules and, more fundamentally, in ensuring the quality of water supplies. Major problems include deteriorated infrastructure, lack of access to capital, limited customer and rate base, inadequate rates, diseconomies of scale, and limited managerial and technical capabilities. Because of these same characteristics, the DWSRF program has not been as successful for small systems, compared to larger systems. Although these systems serve just 9% of the population served by community water systems, the sheer number of small systems has created challenges for policymakers and regulators.

In the earliest SDWA debates, Congress recognized that setting standards based on technologies affordable for large cities could pose problems for small systems. During the reauthorization debate leading up to the 1996 amendments, policymakers gave considerable attention to the question of how to help small systems improve their capacity to comply with SDWA mandates. The 1996 amendments added provisions aimed at achieving this goal, including a requirement that states establish strategies to help systems develop and maintain the technical, financial, and managerial capacity to meet SDWA regulations. Congress also revised provisions on standard-setting (§1412(b)), variances (§1415(e)), and exemptions (§1416) to increase consideration of small system concerns.

Exemptions

The act's exemption provisions are intended to provide compliance flexibility in certain cases. States or EPA may grant *temporary* exemptions from a standard if, due to certain compelling factors (including cost), a system cannot comply on time. For example, all systems are required to comply with the new arsenic standard five years after its promulgation date. An exemption would allow three more years for qualified systems. Small systems (serving 3,300 persons or fewer) may be eligible for up to three additional two-year extensions, for a total exemption duration of nine years (and for a total of up to 14 years to achieve compliance). In the preamble to the arsenic rule published in January 2001, EPA noted that exemptions will be an important tool to help states address the number of systems needing financial assistance to comply with this rule and other SDWA rules (66 *Federal Register* 6988).

However, to grant an exemption, the law requires a state to hold a public hearing and make a finding that the extension will not result in an "unreasonable risk to health." Because of the administrative burden to the states, the act's exemption authority has seldom been used. Approximately 13 states had indicated that they would likely use the exemption process for the arsenic rule, but it appears that many states have not exercised this option.

Small System Variances and Affordability

In contrast to exemptions, variances offer a more permanent form of compliance flexibility for small systems. Since 1996, SDWA has required EPA, when issuing a regulation, to identify technologies that meet the standard and that are affordable for systems that serve populations of 10,000 or fewer. If EPA does not identify affordable "compliance" technologies, then the agency must identify small system "variance" technologies. A variance technology need not meet the standard, but must protect public health. States may grant variances to systems serving 3,300 persons or fewer if a system cannot afford to comply with a rule (through treatment, an alternative source of water, or other restructuring) and if the system installs a variance technology. With EPA approval, states also may grant variances to systems serving between 3,300 and 10,000 people. (Regulations addressing microbial contaminants are not eligible for variances under the statute.)

In 1998, EPA published affordability criteria to establish guidelines for determining whether a regulation is deemed affordable for small systems, and whether small system variances would be available. Under the criteria, EPA evaluates the affordability of a regulation by determining whether the compliance cost would raise the total water cost above 2.5% of annual median household income (MHI) in the three categories of small systems. Using this approach, EPA has determined that affordable compliance technologies are available for every drinking water regulation. Consequently, the agency has not identified any small system variance technologies, and thus, no small system variances are available.

Several recent regulations (such as the revised arsenic and radium standards and the Stage 2 Disinfectants and Disinfection Byproducts Rule) have heightened concern, particularly among rural communities, that EPA has not used the tools Congress provided to help small systems comply with SDWA regulations.

Affordability Criteria Review

Prompted by debate over the revised arsenic standard and its potential cost to small communities, the conference report for EPA's FY2002 appropriations (H.Rept. 107-272) directed EPA to review its affordability criteria and how small system variance programs should be implemented for the arsenic rule. EPA began the review and sought the advice of the EPA's National Drinking Water Advisory Council (NDWAC) and Science Advisory Board (SAB).

After considering recommendations from its affordability work group, the NDWAC reported to EPA in 2003. The council acknowledged the statutory basis for small system variances and recommended changes, but cautioned that "significant practical, logistical, and ethical issues mitigate against the use of variances."[27] The National Rural Water Association, a member of the NDWAC work group, dissented and issued a separate report urging EPA to adopt a safe and affordable variance approach that would make variances available to small communities, as authorized by Congress. The Science Advisory Board concluded that EPA's basic approach was justified on the basis of equity, efficiency and administrative practicality, but recommended ways to improve the criteria. The SAB suggested that EPA consider lowering its affordability threshold, noting that "the national affordability threshold has never been exceeded, but some small water systems appear to have genuinely struggled with costs, suggesting that the 2.5% rule is too high."[28] The SAB also encouraged EPA to develop clear guidelines about when variances should be granted, and recommended that EPA consider measures other than median income to better capture impacts on disadvantaged households.

In March 2006, EPA proposed three options for revising its affordability criteria for determining whether a compliance technology is unaffordable for small systems (71 *Federal Register* 10671). EPA currently assumes that treatment technology costs are affordable to the average household if they do not cause median annual water bills to exceed about $1,000 (this threshold is calculated by taking 2.5% of median household income among small systems). Based on this approach, EPA has determined that affordable technologies are available for all standards. The three proposed options are well below that level: 0.25%, 0.50%, and 0.75%. EPA also requested comment on whether or not the agency should evaluate affordability strictly on a national level, or use a two-step process that would include evaluations of affordability first at the national level and then at the county level. A county level analysis would be performed only when a standard was found to be affordable at the national level. The revised criteria are further intended to address the issue of how to ensure that a variance technology would be protective of public health — an issue that has historically hampered the use of variances.

EPA is evaluating comments on its proposed revisions, and has noted its intention to apply the revised criteria only to the recent Stage 2 DBP Rule and future rules. States could use the criteria to grant small-system variances, on a case-by-case basis, when systems cannot afford to comply with a standard. If these variances become available, it is not clear how often they might be used. A key issue is that variances allow systems to provide lower-quality water in lower-income communities, and this could raise issues for states, communities, and consumers.

Small System Legislation

During the 110th Congress, various bills were introduced to help small water systems comply with the arsenic standard and other drinking water regulations. Legislation has focused on promoting small system compliance through funding assistance and compliance flexibility. The Senate Environment and Public Works Committee reported S. 1933, the Small Community Drinking Water Funding Act (S.Rept. 110-475) which would have directed EPA to establish a small public water system grant program. As mentioned, the committee also reported S. 3617 (S.Rept. 110-509) to authorize increased funding for the DWSRF program, and to create a grant program at EPA for small or economically disadvantaged communities for critical drinking water and water quality projects. Additionally, S. 199 (S.Rept. 110-476) would have increased the authorization of appropriations for grants to Alaska to build water and wastewater systems in rural and Native villages. Among other small-system focused bills, H.R. 2141 would have required primacy states to grant exemptions to eligible small systems for rules covering naturally occurring contaminants, including arsenic, radon, radium, uranium, and several disinfection byproducts. S. 2509, the Small System Safe Drinking Water Act of 2007, proposed to provide more compliance assistance to small communities and prevent the enforcement of certain drinking water regulations unless sufficient funding was available or a variance technology had been identified. S. 2509 also would have required EPA to establish a research program to explore new compliance technologies, and would have established new requirements for EPA's affordability criteria. S. 1429 (S.Rept. 110-242) would have reauthorized SDWA funding for small system technical assistance.

Underground Injection Control and Carbon Sequestration

Most public water systems rely on groundwater as a source of drinking water, and the 1974 Safe Drinking Water Act authorized EPA to regulate the underground injection of fluids (including solids, liquids, and gases) to protect underground sources of drinking water.[29] SDWA §1421 directed EPA to promulgate regulations for state underground injection control (UIC) programs, and mandated that the regulations contain minimum requirements for programs to prevent underground injection that endangers drinking water sources.[30] Section 1422 authorized EPA to delegate primary enforcement authority (primacy) for UIC programs to the states, provided that state programs prohibit anyunderground injection that is not authorized by a state permit.[31] Thirty-three states have assumed primacy for the program, EPA has lead implementation and enforcement authority in 10 states, and authority is shared in the remainder of the states.[32]

The 1974 law specified that the UIC regulations could not interfere with the underground injection of brine from oil and gas production or recovery of oil unless underground sources of drinking water would be affected.[33] In the Energy Policy Act of 2005, the 109th Congress amended SDWA to specify further that the definition of "underground injection" excludes the injection of fluids or propping agents (other than diesel fuels) used in hydraulic fracturing operations related to oil, gas, or geothermal production activities.[34]

The UIC program regulations specify siting, construction, operation, closure, financial responsibility, and other requirements for owners and operators of injection wells. EPA has established five classes of injection wells based on similarity in the fluids injected and activities, as well as common construction, injection depth, design, and operating techniques.

Carbon Sequestration and Storage

In the 110th Congress, underground injection received attention primarily regarding its role as a potential means for sequestering carbon dioxide (CO_2) emissions in geologic formations to control greenhouse gas emissions.

Geologic sequestration (GS) is the process of injecting CO_2 captured from a large stationary source (such as a coal-fired power plant) through a well deep into the earth for long-term storage. Research indicates that numerous geologic formations exist in the United States and worldwide that have the capacity to store large volumes of CO_2. Because coal is responsible for nearly half of the electricity generated worldwide and its use is increasing, carbon capture and storage (CCS) is attracting a growing number of proponents who believe that, with proper site selection and management, geologic sequestration could play an important role in controlling CO_2 emissions.

Although considerable interest has emerged for the rapid, commercial-scale development of carbon sequestration projects, questions exist regarding the long-term safety and effectiveness of sequestration of large volumes of CO_2. Issues include how sequestration activities might affect underground sources of drinking water, what local health and environmental risks could arise from slow leakage or sudden releases of stored gas, and who would have long-term responsibility for water contamination or other damages that might result from sequestration activities.

A key public health and environment issue concerns the potential for stored CO_2 to contaminate underground water supplies or otherwise adversely affect human health and the environment. According to a 2005 report by the United Nations Intergovernmental Panel on Climate Change (IPCC), human and environmental risks potentially could result from leaking injection wells, abandoned wells, or leakage across faults in rock formations and ineffective confining layers. The IPCC report noted that,

> Avoiding or mitigating these impacts will require careful site selection, effective regulatory oversight, an appropriate monitoring program that provides early warning that the storage site is not functioning as anticipated and implementation of remediation methods to stop or control CO_2 releases. Methods to accomplish these are being developed and tested.[35]

Noting that knowledge gaps exist and that more demonstration projects are needed, the IPCC report concluded that, although "more work is needed to improve technologies and decrease uncertainty, there appear to be no insurmountable technical barriers to an increased uptake of geological storage as an effective mitigation option."[36] However, uncertainties and research gaps involving the safety and effectiveness of long-term carbon sequestration, the potential health and environmental impacts, regulatory requirements, and long-term liability all pose hurdles to the rapid deployment of this technology.[37]

Congress has acted on several bills that would facilitate and/or regulate the use of underground injection wells for the purpose of carbon sequestration. Enacted in December 2007, the Energy Independence and Security Act of 2007 (EISA, P.L. 110-140) expands the Department of Energy (DOE) carbon sequestration research and development program. EISA Section 702 requires DOE to conduct at least seven large-volume sequestration tests, in addition to conducting research that promotes the development of sequestration technologies. Section 706 specifies that the injection and sequestration of CO_2 under EISA will be subject to the requirements of the Safe Drinking Water Act, including the act's UIC provisions.

In addition to EISA, several bills in the 110th Congress that contained geologic sequestration provisions would have required sequestration activities to be done in conformance with SDWA requirements. Both S. 3036 and H.R. 6186 proposed to amend SDWA to require EPA to promulgate regulations to manage and facilitate carbon sequestration.[38]

In July 2008, EPA proposed regulations to create a nationally consistent framework for managing the underground injection of CO_2 for geologic sequestration purposes, thus taking a step toward providing certainty to industry and the public about requirements that would apply to this activity.[39] The rule proposes to create a new class of injection wells (Class VI) for geologic sequestration, and establish national requirements that would apply to these injection wells. The proposed rule builds on the existing UIC program, including requirements for well owners and operators to ensure that wells are appropriately located, constructed, tested, monitored, and ultimately closed with proper funding.

EPA's stated regulatory goal is to ensure that permittingregulations are in place to ensure that GS can occur in a safe and effective manner in order to enable commercial-scale CCS projects to move forward. After taking public comment on the rule through December 2008, the agency expects to promulgate a final GS rule under SDWA in December 2010 or 2011. EPA is coordinating with DOE on carbon sequestration research, development, and demonstration activities.

CONGRESSIONAL HEARINGS AND REPORTS, AND DOCUMENTS

U.S. Congress. Senate. Committee on Environment and Public Works. *Water Infrastructure Financing Act.* Report to accompany S. 1400. 109th Cong., 1st sess. December 8, 2005. 52 p. (S.Rept. 109-186).

U.S. Congress. House. Committee on Government Reform. *Public Confidence, Down the Drain: the Federal Role in Ensuring Safe Drinking Water in the District of Columbia.* Hearing, March 5, 2004, 108th Cong., 2nd sess. 268 p. (H.Rept. 108-161).

U.S. Congress. House. Committee on Government Reform. Subcommittee on Energy Policy, Natural Resources and Regulatory Affairs. *EPA Water Enforcement: Are We on the Right Track?* Hearing, October 14, 2003, 108th Cong., 1st sess. 201p. (H.Rept. 108-157).

U.S. Congress. House. Committee on Transportation and Infrastructure. Subcommittee on Water Resources and Environment. *Aging Water Supply Infrastructure.* Hearing, April 28, 2004, 108th Cong., 2nd sess. 78 p. (H.Rept. 108-63).

ADDITIONAL READING

U.S. Environmental Protection Agency. *Drinking Water State Revolving Fund Program: Increasing Impact, 2006 Annual Report.* Office of Water. Report No. EPA 816-R-07-002, June 2007. 44 p. [http://www.epa.gov/safewater/dwsrf/ index.html]

U.S. Environmental Protection Agency. *Providing Safe Drinking Water in America: 2003 National Public Water Systems Compliance Report.* Office of Enforcement and Compliance Assurance. Report No. EPA 305-R-05-002. September 2005. 19 p. plus appendices. [http://www.epa.gov/compliance/resources/reports/accomplishments/ sdwa/sdwacom2003.pdf]

U.S. Environmental Protection Agency. *The Clean Water and Drinking Water Infrastructure Gap Analysis Report.* Office of Water. Report No. EPA 816-R-02-020. September 2002. 50 p.

National Research Council. *Health Implications of Perchlorate Ingestion.* Board on Environmental Studies and Toxicology. National Academies Press. January 2005. 177 p.

End Notes

[1] Title XIV of the Public Health Service Act, as added by P.L. 93-523 and subsequently amended (42 U.S.C. 300f-300j-26).

[2] For purposes of the PWSS program, the term "state" includes 57 states, commonwealths, and territories that have been approved to implement the drinking water program within their jurisdiction. It also includes the Navajo Nation, which received EPA approval to implement its drinking water program in 2000.

[3] U.S. Environmental Protection Agency, *Providing Safe Drinking Water in America: 2003 National Public Water Systems Compliance Report*. Office of Enforcement and Compliance Assurance. Report No. EPA 305-R-05-002. September 2005. 19 p. plus appendices.

[4] U.S. Environmental Protection Agency, *FY2009 Annual Performance Plan and Congressional Justification,* Program Performance and Assessment, p. 795.

[5] For more information, see CRS Report RL31294, *Safeguarding the Nation's Drinking Water: EPA and Congressional Actions*, by Mary Tiemann.

[6] Information on the CCL2 and the rationale behind EPA's regulatory determinations are available at [http://www.epa.gov/safewater/ccl/reg_determine2.html#fr].

[7] January 4, 2007 (72 *Fed. Reg.* 367-398)

[8] For a more detailed discussion, see CRS Report RL31243, *Safe Drinking Water Act: A Summary of the Act and Its Major Requirements*, by Mary Tiemann.

[9] U. S. Senate. *Safe Drinking Water Amendments Act of 1995*, Report of the Committee on Environment and Public Works on S. 1316. S.Rept. 104-169. p. 14. November 7, 1995.

[10] SDWA §1412(b)(6); 42 U.S.C. 300g-1.

[11] Information on these rules can be found at [http://www.epa.gov/safewater/disinfection].

[12] The newly revised Lead and Copper Rule and more information on lead in drinking water are available at [http://www.epa.gov/safewater/lcrmr/index.html].

[13] For further discussion, see CRS Report RS21961, *Perchlorate Contamination of Drinking Water: Regulatory Issues and Legislative Actions*, by Mary Tiemann.

[14] National Research Council, *Health Implications of Perchlorate Ingestion*, Board on Environmental Studies and Toxicology, National Academies Press, January 2005, 177 p.

[15] Michael A. Kelsh et al., "Primary Congenital Hypothyroidism, Newborn Thyroid Function, and Environmental Perchlorate Exposure Among Residents of a Southern California Community," *Journal of Occupational Environmental Medicine*,45(10) p. 1116-1127, October 2003.

[16] Benjamin C. Blount, James L. Pirkle, et al., "Urinary Perchlorate and Thyroid Hormone Levels in Adolescent and Adult Men and Women Living in the United States," Centers for Disease Control and Prevention, *Environmental Health Perspectives*, 114(12), p. 1865-1871, December 2006.

[17] For more information on EDCs and potential health risks, see CRS Report RL31267, *Environmental Exposure to Endocrine Disruptors: What Are the Human Health Risks?*, by Linda Schierow and Eugene H. Buck.

[18] See for example, U.S. Geological Survey, *Pharmaceuticals, Hormones, and Other Organic Wastewater Contaminants in U.S. Streams*, USGS FS-027-02, June 2002.

[19] Aquatic organisms face higher risks of exposure than humans for several reasons. For example, these organisms have continuous exposure, and generally are exposed to higher concentrations of PPCPs in untreated water, compared to treated drinking water.

[20] For further information on PPCPs and related EPA activities, see [http://epa.gov/ppcp].

[21] Information of the CCL3 is available at, [http://www.epa.gov/OGWDW/ccl/ccl3.html].

[22] Association of Metropolitan Water Agencies, *AMWA Discusses Pharmaceuticals in Water Supplies*, March 11, 2008, [http://www.amwa.net].

[23] See also CRS Report RS22037, *Drinking Water State Revolving Fund: Program Overview and Issues*, by Mary Tiemann. For information on other assistance programs, see CRS Report RL30478, *Federally Supported Water Supply and Wastewater Treatment Programs*.

[24] U.S. Environmental Protection Agency, *Drinking Water Infrastructure Needs Survey and Assessment: Third Report to Congress*, June 2005, EPA 816-R-05-001, available at [http://www.epa.gov/safewater/needssurvey/index.html].

[25] U.S. Environmental Protection Agency, *The Clean Water and Drinking Water Infrastructure Gap Analysis Report*, Report No. EPA 816-R-02-020, September 2002, 50 p.

[26] For further discussion of infrastructure issues, see CRS Report RL31116, *Water Infrastructure Needs and Investment: Review and Analysis of Key Issues*, by Claudia Copeland and Mary Tiemann.

[27] U.S. Environmental Protection Agency, *Small Drinking Water Systems Variances: Revision of Existing National-Level Affordability Methodology and Methodology to Identify Variance Technologies that Are Protective of Public Health*, (71 Fed. Reg. 10671), March 2, 2006, p. 10657.

[28] U.S. Environmental Protection Agency, Science Advisory Board, *Affordability Criteria for Small Drinking Water Systems: An EPA Science Advisory Board Report*, 2002, p. 4. The SAB report is available at [http://www.epa.gov/safewater/pws/affordability.html].

[29] Underground injection control provisions are contained in SDWA §1421 - §1426; 42 U.S.C. 300h - 300h-5.

[30] § 1421(d)(2) states that

> underground injection endangers drinking water sources if such injection may result in the presence in underground water which supplies or can reasonably be expected to supply any public water system of any contaminant, and if the presence of such contaminant may result in such system's not complying with any national primary drinking water regulation or mayotherwise adversely affect the health of persons.

[31] P.L. 93-523, SDWA §1421 (42 U.S.C. 300h).

[32] To receive primacy, a state, territory, or Indian tribe must demonstrate to EPA that its UIC program is at least as stringent as the federal standards; the state, territory, or tribal UIC requirements may be more stringent than the federal requirements. For Class II (oil and gas) wells, states must demonstrate that their programs are effective in preventing pollution of underground sources of drinking water (USDWs).

[33] SDWA §1421(d) specifies that "underground injection" does not include the underground injection of natural gas for storage purposes. In legislative history, Congress explained that the natural gas exclusion applies only to "natural gas as it is commonly defined" and "not to other injections of matter in a gaseous state." U.S. House of Representatives, H.Rept. 961348, 1980, USCCAN, p. 6080.

[34] P.L. 109-58, H.R. 6, Section 322,amended SDWA section 1421(d).

[35] United Nations Intergovernmental Panel on Climate Change, 2005, IPCC Special Report on Carbon Dioxide Capture and Storage, p. 197.

[36] Ibid, p. 198.

[37] Commercial-scale deployment of CCS faces a range of technical, legal, economic, regulatory, and public policy issues. Capturing carbon and preparing it for transport and storage are generally considered the most economically and technologically challenging aspects of CCS, and no commercial technology to capture these emissions is currently available for large-scale coal-fired power plants. Moreover, carbon capture technologies would markedly increase the cost of electricity generation. Consequently, few companies may be inclined or able to install such technology unless they are required to do so, either by regulation or by a carbon price. For further discussion see CRS Report RL34621, *Capturing CO2 from Coal-Fired Power Plants: Challenges for a Comprehensive Strategy*, by Peter Folger, Larry Parker, and Deborah D. Stein.

[38] For a detailed discussion of geologic sequestration and related legislation, see CRS Report RL33801, *Carbon Capture and Sequestration (CCS)*, and CRS Report RL34218, *Underground Carbon Dioxide Storage: Frequently Asked Questions*, both by Peter Folger.

[39] U.S. Environmental Protection Agency, *Federal Requirements Under the Underground Injection Control (UIC) Program for Carbon Dioxide (CO2) Geologic Sequestration (GS) Wells*, Proposed Rule, 73 *Fed. Reg.* 43491-43541, July 25, 2008.

In: Safeguarding the Nation's Drinking Water ISBN: 978-1-60741-234-2
Editors: Juan A. Schrock

Chapter 2

SAFEGUARDING THE NATION'S DRINKING WATER: EPA AND CONGRESSIONAL ACTIONS*

Mary Tiemann

SUMMARY

The events of September 11, 2001, focused heightened attention on the security status of the nation's drinking water supplies and the vulnerability of this critical infrastructure sector to attack. Congress since has enacted security requirements for public water systems and has provided funding for vulnerability assessments, emergency planning, and drinking water research. The Environmental Protection Agency (EPA), the lead federal agency for the water sector, has worked with water utilities, state and local governments, and federal agencies to improve the drinking water security.

The Public Health Security and Bioterrorism Preparedness and Response Act of 2002 (P.L. 107-188) amended the Safe Drinking Water Act to require some 8,400 community water systems to assess vulnerabilities and prepare emergency response plans. It authorized funding for these activities and for emergency grants to states and utilities, and it directed EPA to review methods

---

* This is an edited, reformatted and augmented version of a CRS Report for Congress, Report RL31294, dated June 2, 2008.

to prevent, detect, and respond to threats to water safety and infrastructure security. The act did not require water systems to make security upgrades to address potential vulnerabilities. Since FY2002, Congress has appropriated funds annually for EPA to work with states and the water sector to improve the security of drinking water supplies.

In the Homeland Security Act of 2002 (P.L. 107-296), Congress created a Department of Homeland Security (DHS) and gave the DHS responsibility for assessing and protecting the nation's critical infrastructures. However, the act did not transfer EPA's water security functions, and the 2003 Homeland Security Presidential Directive (HSPD-7) affirmed EPA's lead role in protecting the water infrastructure. Under this directive, EPA has responsibility for developing and providing tools and training on improving security to roughly 53,000 community water systems and 16,000 municipal wastewater treatment facilities.

In the 109th Congress, several bills, including a reported bill, S. 2145, proposed to expand water security requirements for certain high-risk water systems. The Department of Homeland Security FY2007 appropriations act (P.L. 109-295) authorized the DHS to regulate for three years high-risk chemical facilities, but the law excluded from coverage drinking water and wastewater treatment facilities.

Although EPA, states, localities, and water utilities have taken steps to address security concerns, the security of the nation's water supplies continues to attract congressional attention. Issues receiving attention have included the status of efforts by the water sector to improve security, whether to increase federal requirements, funding needs for water systems to make security improvements, the relative roles and responsibilities of EPA and DHS regarding the water sector, and the status of research and development of technologies to help water systems detect and address potential biological and chemical contaminants. This chapter reviews governmental and water utility efforts to improve drinking water security.

BACKGROUND

Ensuring the security of the nations' drinking water supplies poses a substantial challenge, partly because the number of water systems is very large and also because the responsibility for protecting drinking water safety is shared among federal, state and local governments and utilities. Nationwide, there are some 158,000 public water systems, and these systems range greatly in size, serving from as few as 25 persons to more than 1 million persons. Roughly 53,000 of all public water systems are community water systems (CWSs) that serve the same residences year-round.[1] These 53,000 systems provide water to approximately 282 million people. Nearly 400 community systems serve more than 100,000 people and provide water to nearly half of the total population served. Because water supplies support many uses (from drinking water to fire suppression), their disruption could have significant impacts.[2]

A 1996 executive order on critical infrastructure protection (E. O. 13010), included water supply systems as one of eight national infrastructures vital to the security of the United States.[3] In 1997, the President's Commission on Critical Infrastructure Protection, created by the executive order, issued a report on the vulnerabilities of these infrastructure sectors and strategies for protecting them. The Commission identified three attributes crucial to water supply users: water must be available on demand, it must be delivered at sufficient pressure, and it must be safe for use.[4] Actions affecting any of these factors could be debilitating for the infrastructure and also for the communities that depend on it.

Major threats to water supplies include physical destruction of facilities or distribution systems, biological or chemical contamination of supplies, and cyber attacks. The 1997 Commission found that drinking water systems had inadequate protection against the threat of chemical or biological contamination, and that technology was insufficient to allow detection, identification, measurement, and treatment of highly toxic, waterborne contaminants. Water utilities were also found to be vulnerable to cyber attacks as they rely increasingly on computers to control water flow and pressure. Information sharing was identified as the most immediate need, whereas warning and analytical capabilities and research and development all were found to be insufficient.

In response to these findings and related developments, President Clinton, in 1998, issued Presidential Decision Directive (PDD) 63 on critical infrastructure protection. Under this directive, a public/private partnership was established to put in place prevention, response, and recovery measures to ensure the security of the nation's critical infrastructures against criminal or terrorist attacks. PDD-63 designated the Environmental Protection Agency (EPA) as the lead federal agency for the water supply sector, and EPA appointed the Association of Metropolitan Water Agencies (AMWA) to coordinate the water sector. However, before September 11, 2001, the main focus of PDD-63 efforts for all critical infrastructure sectors was on cybersecurity. Subsequently, the breadth and depth of efforts to protect the nation's critical infrastructures have expanded significantly.

EPA Efforts To Increase Drinking Water Security

In 2003, President Bush issued Homeland Security Presidential Directive 7 (HPSD-7), which affirmed EPA as the lead federal agency for coordinating the protection of the nation's critical infrastructure for the water sector. Under this directive, EPA is responsible for developing and providing tools and training on improving security to roughly 53,000 community water systems and 16,000 municipal wastewater treatment facilities.

To carry out its water sector responsibilities, EPA established a Water Security Division within the Office of Ground Water and Drinking Water. This Division works with drinking water and wastewater utilities, states, tribes, and other stakeholders to improve the security of these utilities and improve their ability to respond to security threats and breaches. Among its responsibilities and activities, the Water Security Division provides security and antiterrorism-related technical assistance and trainingto the water sector. Although the Water Security Division was established in 2003, the Office of Water had provided assistance to its stakeholders for a number of years.

Security-related activities undertaken by EPA and the water sector have fallen into five general categories, including (1) establishing an information center for drinking water alerts or incidents, (2) developing vulnerability assessment tools, (3) identifying actions to minimize vulnerabilities, (4) revising emergency operations plans, and (5) supporting research on biological and chemical contaminants considered to be potential weapons of mass destruction. Several key government and private sector efforts are reviewed below.

Information Sharing and Analysis

One goal of PDD-63 in 1998 was to establish an Information Sharing and Analysis Center (ISAC) for each critical infrastructure sector. With assistance from EPA and the Federal Bureau of Investigation, the Association of Metropolitan Water Agencies has led the effort to develop and implement an

ISAC for water utilities. The WaterISAC provides a secure, Web-based communications link between the water sector and federal homeland security, law enforcement, intelligence, environmental, and public health agencies. Thissystemgathers, analyzes, and disseminates threat information specific to the water sector, and is a comprehensive source of security and disaster preparedness information for drinking water and wastewater utilities.[5] The WaterISAC is supported by membership fees and EPA grants. It has been supplemented with the Water Security Channel (WaterSC), which provides free email notification of water security alerts and other information issued by federal government agencies.

Tools and Technical Assistance

EPA has worked with its water sector partners to provide practical tools and technical assistance to utilities on a wide range of security matters. In 2000, the American Water Works Association Research Foundation (AWWARF) and the Sandia National Laboratories, with EPA support, initiated a project to develop a methodology for utilities to use to assess their vulnerabilities and develop plans to minimize identified risks. The project was expedited after September 11, 2001, and completed in November 2001, and many large water systems used this risk assessment methodology to conduct vulnerability assessments. States and drinking water organizations, in collaboration with EPA, developed additional vulnerability assessment tools, with a particular focus on the needs of smaller communities. To help cover the costs of conducting vulnerability assessments and preparing emergency response plans, EPA awarded a total of $51 million in grants to community water systems that serve more than 100,000 persons.

While direct grants have not been available for smaller water systems, a considerable amount of technical assistance has been aimed at helping these systems, which typically may have less capacity to address security concerns. EPA has used "train-the-trainer"grants to build a pool of environmental professionals that has provided training and technical assistance to water systems serving fewer than 50,000 people.[6] In addition, on-site assistance for vulnerability assessment and emergency response planning was made available to small and medium wastewater utilities at no cost through the Wastewater Operator Training Program.

Technical assistance also has been provided through numerous guidance documents designed to help water utilities address a range of security concerns.[7] A key product is the *Response Protocol Toolbox: Planning for and Responding to Contamination Threats to Drinking Water Systems*, which is intended to help public water systems respond to contamination threats and incidents. The "toolbox" includes separate modules that address water utility planning, contamination threat management, site characterization and investigation, water sample analysis, public heath response, and remediation and recovery.[8] Paralleling the toolbox, EPA developed response guidelines to help water utilities, emergency responders, and other officials during the management of an ongoing contamination threat or incident.

As a sector, drinking water utilities acted relatively quickly to assess vulnerabilities, upgrade emergency response plans, and take some initial steps to improve security of this critical infrastructure. The Office of Homeland Security's 2003 *National Strategy for the Physical Protection of Critical Infrastructure and Key Assets* reported that the water sector had taken great strides to protect its critical facilities and systems, and had focused on categories of possible attacks that could have the greatest health or economic consequences.[9] However, the report noted that the water sector needed better threat information to prioritize investments on security measures. It also reiterated the need for research and development of new monitoring and analytic capabilities to enhance detection of biological, chemical, and radiological contaminants that could be introduced to the water supply.

Research

EPA has participated in various research and development projects related to water security, including research to evaluate the ability of drinking water treatment systems to remove and inactivate biological and chemical agents. The agency also has supported research efforts to determine the fate and transport of contaminants within rivers and streams and within water treatment plants and distribution systems, and to develop biodetectors for detecting and quantifying biological contaminants in drinking water supplies.

To coordinate and oversee research involving prevention and response to terrorist attacks, EPA's Office of Research and Development established the National Homeland Security Research Center (NHSRC) in 2002. The Center's key areas of research involve water infrastructure protection, decontamination and consequence management, and threat and consequence assessment.

In 2004, the NHSRC's Water Infrastructure Protection Division and the Office of Water's Water Security Division developed a Water Security

Research and Technical Support Action Plan to define a specific program of research and technical support for protecting drinking water and wastewater facilities from terrorist threats and attacks. The action plan, which was reviewed by the National Research Council (NRC), addressed drinking water supply, water treatment, finished water storage, and distribution systems. It identified major research needs in the following areas:

- protecting physical and cyber infrastructure;
- identifying drinking water contaminants;
- improving monitoring systems and analytical methods for drinking water;
- containing, treating, decontaminating, and disposing of contaminated water and material;
- contingency planning;
- addressing infrastructure interdependencies;
- risk assessment and communication; and
- protecting wastewater treatment and collection systems.[10]

A key concern the NRC expressed regarding the action plan was that it did not discuss the financial resources that would be required to complete the proposed projects and to implement countermeasures needed to improve water security. The NRC recommended that EPA try to quantify the costs and benefits associated with the research and technical support projects. The NRC further noted that more emphasis was needed on communicating the value of water and increased security, because water rate increases would likelybe needed to generate the resources needed to implement counter measures.[11]

In a progress report on the action plan, EPA reported that more than 100 projects had been initiated to address the needs listed in the plan.[12] Projects have included a review of early warning systems, a tracer studies guide for use by water utilities, a treatability guide for biological contaminants in water, a review of emerging detection technologies for water contaminants, a review of the impacts of biological toxins on water systems, and performance verifications of the effectiveness of monitoring, treatment, and decontamination technologies. The Water Infrastructure Protection Division has lead responsibility for much of this research and has been producing tools, guides, and other products for use by water utility operators, public health officials, and emergency responders.

For several years, EPA's major water security research effort has involved the Water Security Initiative (formerly called the WaterSentinel Initiative), which is a demonstration project to develop a model contamination warning system for drinking water systems. The goal of the initiative is to establish pilot early warning systems through intensive water monitoring and surveillance in selected cities. As noted below, Congress has given partial support to the initiative. Using the resources available, EPA published interim guidance for the Water Security Initiative in May 2007, and established the first contaminant warning system pilot in July 2007. The agency plans to award grants for additional pilot projects in March 2008.

FUNDING FOR DRINKING WATER SECURITY ACTIVITIES

Since 2001, Congress has provided funds annually to EPA to improve the security of public water supplies. The *Emergency Supplemental Appropriations Act for FY2002* (P.L. 107-117) provided EPA with $175.6 million for emergency expenses to respond to the September 11 attacks and to support counterterrorism activities. The accompanying conference report, H.Rept. 107-350, specified that approximately $90 million was for improving security at EPA laboratories, performing drinking water vulnerability assessments, and anthrax decontamination activities. Another $5 million was for state grants for counterterrorism coordinators to work with EPA and water utilities in assessing drinking water safety. Congress has continued to provide roughly $5 million for these state grants each year.

During **FY2002**, EPA allocated roughly $89 million of the amount provided in the emergency supplemental appropriation to support security enhancements at the nation's drinking water systems. Of this amount, EPA targeted approximately $80 million to: (1) provide grants to the largest drinking water systems to conduct vulnerability assessments and enhance emergency response plans; (2) provide technical assistance on vulnerability assessments and emergency response plans to small and medium drinking water systems; and (3) refine security-related detection, monitoring, and treatment tools. Another $4 million was used to accelerate the development and testing of counterterrorism tools, support vulnerability assessment training, provide technical assistance, and conduct and implement research on redesign and detection for collection and treatment systems. EPA also used funds to develop tools and provide training for medium and small drinking water systems to assess vulnerabilities and develop emergency response plans. In addition, EPA allocated $5 million to the states to support homeland security coordination work involving EPA and drinking water utilities.

EPA awarded approximately $51 million in water security grants to the community water systems that serve more than 100,000 individuals. Grants were made to roughly 400 publicly and privately owned community water systems for as much as $115,000 per grant. Utilities were able to use their grants to develop vulnerability assessments, emergency response plans, and security enhancement plans and designs. Utilities also could use grant funds for in-house or contractor support; however, funds could not be used for physical improvements.

Although these grants were made only to large systems, EPA has worked with states and utilities to help meet the security needs of small and medium-sized drinking water systems. EPA provided roughly $20 million of FY2002 supplemental funds directly to the states for technical assistance and training for drinking water systems serving fewer than 100,000 people.

For **FY2003**, EPA requested $16.9 million to assist small and medium-sized systems with vulnerability assessments and emergency response plans, and $5 million in grants to states to support homeland security coordination. The Consolidated Appropriations Resolution for FY2003 (P.L. 108-7), provided this amount. It also contained several drinking water security earmarks, including $2 million for the National Rural Water Association to help small water systems conduct vulnerability assessments and $1 million for theAmerican Water Works Association for water security training activities.

As requested for **FY2004**, EPA received approximately $32 million for critical water infrastructure protection, including $5 million for state homeland security grants in P.L. 108-199. This funding supported states' efforts to work with water and wastewater systems to develop and enhance emergency operations plans; conduct training in the implementation of remedial plans in small systems; and develop detection, monitoring and treatment technology to enhance water security. EPA used funds to assist the nearly 8,000 community water systems that serve water to populations between 3,300 and 100,000 and are subject to the Bioterrorism Act. P.L. 108-199 also included $2 million for the Water ISAC to gather, analyze and disseminate sensitive security information to water and wastewater systems.

For **FY2005**, EPA requested $5 million for state water security grants and $6.1 million for other critical infrastructure protection efforts. EPA's budget justification explained that the $21.3 million reduction reflected a shift in priorities from assistance and training on vulnerability assessments. (Under the Bioterrorism Act, community water systems were required to complete vulnerability assessments by June 30, 2004.) Congress provided the requested amount in the Consolidated Appropriations Act, FY2005 (P.L. 108-447). As in FY2004, the appropriated amount included $2 million for the Water ISAC.

In the **FY2006** budget request, the President again requested $5 million for state water security grants. The President also requested $44 million to launch the Water Sentinel Initiative, a demonstration project to develop a model contamination warning system, now called the Water Security Initiative. EPA initiated this project to meet its responsibilities under Homeland Security Presidential Directive (HSPD) 9, which directed EPA to develop a surveillance and monitoring program to provide early detection in the event of a terrorist attack contaminating water. The goal of the initiative is to establish pilot early warning systems through intensive water monitoring and surveillance for certain chemical and biologicalcontaminants in five cities. Further responding to HSPD-9, EPA proposed to form a laboratory network to support the monitoring and response requirements of the surveillance program.[13]

In EPA's FY2006 appropriations act (P.L. 109-54), Congress provided $8.1 million (after rescissions) of the $44 million requested for the Water Security Initiative. In recommending a large reduction, the House Appropriations Committee recommended that EPA develop clear goals for the initiative, seek the advice of the Science Advisory Board, and justify the request more clearly in the budget request for FY2007.[14] Congress also provided $5 million ($4.93 after rescissions) for state water security grants.

The **FY2007** budget request included $4.95 million for state water security grants. Additionally, the request again included a significant amount, $41.7 million ($33.6 million more than Congress provided for FY2006), for the Water Security Initiative (WSI). In its justification for the request, the agency noted that the program is an essential component of its water security activities, and explained that its purpose is to demonstrate an effective contamination warning system that could be used by drinking water utilities of various sizes.

The EPA FY2007 funding bill, H.R. 5386, as passed by the House, would have provided $16.7 million (or $25 million less than requested) for the WSI. The Senate Appropriations Committee (S.Rept. 109-275) recommended $18.13 million, which was $23.6 million less than requested but $10 million above the FY2006 enacted level. The House Appropriations Committee report for H.R. 5386 (H.Rept. 109-465) stated that the committee's recommended funding level for the initiative included money for one additional WSI pilot project, which should be located in a metropolitan area that is highly vulnerable from a homeland security threat perspective. Congress did not complete action on this appropriations bill. The final continuing appropriations resolution for FY2007 (P.L. 110-5, H.J.Res. 20) generally funded EPA at the FY2006 level, but EPA used its discretion and allocated $27 million for the Water Security Initiative.

For **FY2008**, EPA requested $25.6 million to support its water sector responsibilities to protect critical water infrastructure, including $21.88 million for the WSI. The agency proposed to use the requested funds to support the existing pilot and to establish additional pilots, with a goal of having all planned pilots under way by 2008. EPA also proposed to continue providing special assistance to high-priority drinking water systems under the Water Alliance for Threat Reduction program. The goal of this assistance is to ensure that water utilities have tools and information to prevent, detect, respond to, and recover from terrorist attacks, other intentional acts, and natural disasters. The request again included $4.95 million for state water security grants.

The House approved EPA's appropriations bill for FY2008, H.R. 2643 in June 2007. The committee report for H.R. 2643 (H.Rept. 110-187, p. 98) recommended reducing the amount requested for the WSI by $3.88 million. The House Appropriations Committee explained that late action on the 2007 budget had delayed the obligation of funds for pilot projects until FY2008. In the Senate, the Appropriations Committee recommended a $10 million decrease for the initiative (S. 1696, S.Rept. 110-91, p. 52), also noting that EPA plans to carry forward a large balance of unobligated funds. Both the House and Senate committees urged EPA to be prepared to report to the committees on the status and accomplishments of the WSI pilot projects. After applying the 1.56% rescission, the Consolidated Appropriations Act for FY2008 (P.L. 110-161) provided $11.7 million for the WSI and $4.87 million for state grants.

For **FY2009**, EPA has requested $21.4 million for the Water Security Initiative, $1.3 million for the Water Alliance for Threat Reduction, and $4.9 million for state grants. EPA proposes to use WSI funds to support two more pilots, for a cumulative total of five pilots. Additionally, EPA proposes to continue working with sector partners to establish the Water Laboratory Alliance (WLA). The alliance would provide a network of laboratories that would be able to help confirm and speed responses to intentional contamination events. EPA further proposes to use FY2009 funds to continue to support research to develop methods to detect chemical and biological contaminants and to evaluate event detection software. The agency estimates that 90% of high-concern contaminants lack validated analytical methods.[15]

Thus far, Congress has not provided funding in EPA appropriations for grants to public water systems specifically for making security improvements. However, EPA has identified numerous security measures that are eligible for funding through the Drinking Water State Revolving Fund (DWSRF) program.[16] Eligible measures include making facility improvements, such as adding fencing, cameras, and lighting; securing chemical and fuel storage; hiring guards; and adopting enhanced filtration and disinfection treatment. Congress has provided approximately $845 million annually for this program in recent years. However, it is uncertain how readily funds might become available for security measures, as the key purpose of the DWSRF is to facilitate compliance with federal drinking water regulations, and competition for these funds can be considerable.

Another potential source of funding for community water systems is the State Homeland Security Grant Program, administered by the Department of Homeland Security (DHS). Congress provided $525 million for FY2007 and $950 million for FY2008 for this program, which provides assistance to states to detect, prevent, and respond to terrorist attacks. States are required to allocate 80% of the grant funds received under this program to localities, in accordance with their approved homeland security plans. Funds may be used for homeland security-related training and for protecting critical infrastructure, including making physical security improvements. Local public works agencies, including water districts, are eligible to receive funding from the state; however, most of these funds have been used to support first responders.[17]

BIOTERRORISM PREPAREDNESS AND RESPONSE ACT OF 2002

In 2002, two major laws were enacted that address the security of the nation's critical infrastructure. While the Homeland Security Act of 2002 broadly addressed critical infrastructure protection, the Bioterrorism Preparedness and Response Act of 2002 specifically aimed at improving the security of drinking water supplies.

Title IV of the *Public Health Security and Bioterrorism Preparedness and Response Act of 2002* (P.L. 107-188, 42 U.S.C. 300i) amended the Safe Drinking Water Act (SDWA) to require community water systems serving more than 3,300 individuals to conduct an assessment of their system's vulnerability to terrorist attacks or other intentional acts to disrupt the provision ofa safe and reliable drinking water supply. These water utilities were required to certify that they had conducted a vulnerability assessment and to submit a copy of the assessment to EPA. The act also required the utilities to prepare or revise emergency response plans incorporating the results of the vulnerability assessments no later than six months after completing them. (**Table 1** outlines the schedule for the roughly 8,400 water systems that were required to submit vulnerability assessments to EPA and complete emergency response plans.) As required, EPA issued guidance on conducting vulnerability assessments, preparing emergency response plans, and addressing threats to assist smaller water systems that were not covered by the Bioterrorism Act.[18] EPA reports that the large- and medium-sized water systems achieved 100% compliance with the act's requirements, while smaller systems achieved nearly 100% compliance.

The act exempted the contents of the vulnerability assessments from disclosure under the Freedom of Information Act (except for information contained in the certification that identified the system and the date of the certification). As required by the Bioterrorism Act, EPA developed protocols to protect the vulnerability assessments from unauthorized disclosure. The act provides for civil and criminal penalties for inappropriate disclosure of information by government officials.

The Bioterrorism Act authorized $160 million for FY2002, and such sums as may be necessary for FY2003-FY2005, to provide financial assistance to community water systems to conduct vulnerability assessments, to prepare response plans, and for expenses and contracts to address basic security enhancements and significant threats. (Security enhancements could include purchase and installation of intruder detection equipment and lighting, enhancing security of automated systems, personnel training and security screening of employees or contractors, etc. Funding could not be used for personnel costs, plant operations, monitoring or maintenance.)

For grants to states and water systems to assist in responding to emergency situations, the act authorized $35 million for FY2002, and such sums as may be necessary thereafter. Finally, the act authorized $15 million for FY2002, and such sums as may be necessary for FY2003 through FY2005, for EPA to review methods by which terrorists or others could disrupt the provision of safe water supplies, and methods for preventing, detecting, and responding to such disruptions.

Table 1. Community Water System Requirements under the Bioterrorism Act

System size by population served (est. no. of systems)	Date for completing vulnerability assessments	Date for completing emergency response plans
100,000 or more (approx. 400)	March 31, 2003	September 30, 2003
50,000 - 99,999 (approx. 460)	December 31, 2003	June 30, 2004
3,301 - 49,999 (approx. 7,500)	June 30, 2004	December 31, 2004

EPA and DHS Water Infrastructure Security Roles

The *Homeland Security Act of 2002* (P.L. 107-296) combined the functions of all or parts of 22 federal agencies and departments into a new Department of Homeland Security (DHS). The act gave key responsibility for critical infrastructure protection to the DHS, but did not transfer EPA water security functions to the new department.[19]

With the establishment of DHS, which has overall responsibility for critical infrastructure vulnerability assessment and protection, the relative roles and responsibilities of EPA and DHS were not clear. In late 2003, the White House issued Homeland Security Presidential Directive 7 (HSPD-7), which superseded PDD-63. This Directive established national policy and outlined the roles and responsibilities of federal departments and agencies regarding critical infrastructure protection. It identified EPA as the federal agency (Sector-Specific Agency (SSA)) with lead responsibilities for ensuring the protection of the water infrastructure sector from terrorist attacks or sabotage. Under HSPD-7, DHS is responsible for overall coordination and integration of national critical infrastructure protection efforts by federal, state, and local governments and the private sector, whereas EPA is responsible for developing and providing water security tools and training for the nation's community water systems and municipal wastewater treatment facilities.

Some additional articulation of EPA's role was provided by HSPD-9, which established a national policy to defend the nation's water, agriculture, and food systems against terrorist attacks, major disasters, and other emergencies. This directive instructed EPA to develop a comprehensive surveillance and monitoring program to provide early detection of contaminants in water systems. HSPD-9 also directed EPA to develop an integrated network of water quality laboratories to support the surveillance program. EPA has pursued these responsibilities through its Water Security Initiative and Water Alliance for Threat Reduction program.

As both EPA and DHS have responsibilities for protecting critical water infrastructure, the potential for overlap and duplication is perhaps unavoidable. EPA and DHS, for example, have separate communications and information-sharing networks, and have had different policy advisory groups. In recognition of the need for a coordinating entity for the water sector, the major drinking water and wastewater organizations established the Water Sector Coordinating Council (WSCC) in 2004. This council is associated primarily with DHS, but coordinates with both agencies. In 2005, EPA and DHS facilitated the formation of a parallel Water Sector Government Coordinating Council (GCC) to enable interagency and cross-jurisdictional coordination. Members include DHS, EPA, the U.S. Army Corps of Engineers, the Bureau of Reclamation, the Federal Energy Regulatory Commission, the Centers for Disease Control and Prevention, state water and wastewater associations, and others. Chaired by EPA and co-chaired by DHS, the GCC coordinates strategies, activities, policies, and communication across government entities. The WSCC and GCC work together to coordinate critical infrastructure protection activities within the water sector.

A key requirement for DHS under HSPD-7 was to develop a national strategy to protect all critical infrastructure and key resources (CI/KR). In June 2006, DHS issued the National Infrastructure Protection Plan (NIPP). The NIPP is intended to provide a unifying structure for integratingCI/KR protection efforts into one national program. More specifically, it describes processes for 1) setting security goals; 2) identifying key assets; 3) assessing risks; 4) prioritizing assets as a basis for allocating resources; 5) implementing protection programs; and 6) measuring the effectiveness of CI/KR protection efforts.[20]

The NIPP called for each sector's lead agency to work with its sector to develop Sector Specific Plans (SSPs) consistent with the NIPP. All 17 SSPs were released in May 2007, and EPA's Water SSP is one of seven plans that are publicly available.[21] Taken together, the NIPP and the SSP are intended to provide the structure necessary to coordinate and synchronize activities under various laws, presidential directives, strategies, and initiatives into a unified national approach to protecting critical infrastructure. The goal of the Water

SSP is to develop the sector's strategy and programs to protect CI/KR assets, identify priorities based on risk analysis, describe the resources needed to protect CI/KR, track progress, identify gaps, establish research and development priorities, and work with DHS to continuously improve the NIPP. The SSP also helps define the roles and responsibilities of EPA (as the Water Sector SSA) and other sector partners. Overall, the Water SSP is intended to assist drinking water and wastewater utilities to be better prepared to prevent, detect, respond to, and recover from terrorist attacks, other intentional acts, natural disasters, and other hazards. The plan's overarching goals are to 1) sustain protection of public health and the environment; 2) recognize and reduce risks in the water sector; 3) maintain a resilient infrastructure; and 4) increase communication, outreach, and public confidence.

Issues and Legislation

The 109th Congress considered a variety of bills specifically addressing the security of public water supplies, but no legislation was enacted. Some interest focused on implementation of the drinking water security provisions of the Bioterrorism Preparedness Act. S. 1426, for example, would have reauthorized appropriations for SDWA section 1434 (addressing contaminant prevention, detection, and response), and would have required EPA to report to Congress on progress and problems encountered in implementing these provisions.

A key water sector issue has involved the status and adequacy of public and private efforts to improve the security of public water systems. Because of actions on the part of drinking water utilities, EPA, and the Congress, vulnerability assessment and planning efforts in the water sector have proceeded more rapidly than those in certain other sectors (such as chemical facilities); however, it is unclear how much has been done within the sector to implement security upgrades. Although the Bioterrorism Preparedness and Response Act required community water systems to conduct vulnerability assessments and prepare emergency response plans, it did not require systems to make security upgrades to address any identified vulnerabilities.[22]

Chemical Facility Security

An issue relevant to the water sector that has received significant legislative attention for several years involves the security of chemical facilities, including certain water utilities that are located where a terrorist attack could cause harm to nearby populations. For water utilities, the primary concern is the onsite storage hazardous, gaseous chemicals (such as chlorine) that pose potential risks to local communities if released.

The 109th Congress

Several bills in the 109th Congress targeted such high-consequence facilities within the water sector. S. 2855 proposed to amend SDWA to require community water systems to replace hazardous, gaseous chemicals with inherently safer technologies (IST), for example, switching from the use of chlorine gas to liquid chlorine). S. 2855 also would have required EPA to provide grants to high-consequence facilities for use in paying capital expenditures neededto make the transition to the use of IST. S. 2781 and S. 1995 proposed to amend the Clean Water Act to address security at wastewater treatment facilities.

The Senate Committee on Environment and Public Works reported a wastewater treatment facilities bill, S. 2781 (S.Rept. 109-345), which did not contain IST requirements but did authorize grants to wastewater treatment facilities for security-related efforts, including conducting vulnerability assessments, preparing site security plans, and making security upgrades. No further action was taken on this bill, and identical legislation has been introduced in the 110th Congress (S. 1303).

Broader chemical facility security bills also were offered in the 109th Congress that had implications for water utilities. These bills generally would have authorized the Secretary of the Department of Homeland Securityto regulate chemical facilities, including water treatment plants, that pose certain risks. S. 2145, as reported by the Senate Homeland Securityand Governmental Affairs Committee (S.Rept. 109-332), and its companion bill, H.R. 4999, would have directed the Secretary of DHS to issue rules designating which chemical facilities would be subject to regulation, and establish security performance standards such facilities. Under these bills, facilities would have been required to submit to the DHS vulnerability assessments, security plans, and emergency response plans for terrorist incidents. H.R. 5695 (H.Rept. 109-707) shared several similarities with S. 2145 but would have exempted water facilities covered by the legislation from redundant requirements (such as conducting vulnerability assessments), unless the DHS determined that more stringent security requirements were needed. S. 2486 proposed to cover a wider range of facilities and establish a general duty to ensure that facilities would be designed, operated, and maintained in safe manner; the bill defined this obligation to include use of inherently safer technology to the maximum extent practicable. H.R. 1562 would have required consultation between DHS and EPA, and would have imposed stronger security and emergency planning measures, rather than requiring changes in technology. H.R. 2237 would have expanded EPA's existing authority to oversee chemical facilities but would have required consultation with DHS.

Several local government and water organizations, including the American Water Works Association (AWWA), soughtexemptions from S. 2145 and other bills that proposed to give the DHS authority to regulate water utilities that use hazardous chemicals (such as chlorine gas). These stakeholders argued that EPA already has an established water security program and has been designated the lead agency for water infrastructure security. In general, the AWWA has opposed legislation that would require water utilities to switch treatment processes without considering specific utility circumstances and local water and climate characteristics. Others have argued that mandating the adoption of safer technologies is warranted because of the potential risk that hazardous chemicals, particularly gaseous chlorine, may pose to communities.

The 109th Congress included a chemical facility security provision in the DHS FY2007 appropriations act (P.L. 109-295, Section 550). The provision authorized the department to regulate, for three years, high-risk chemical facilities, *excluding* drinking water and wastewater treatment facilities and facilities in ports. The act directed DHS to establish risk-based security performance standards for designated chemical facilities, and to require these facilities to preparevulnerability assessments and security plans.

To implement this provision, the DHS promulgated regulations in June 2007 that established Chemical Facility Anti-Terrorism Standards (CFATS). These standards require chemical facilities to report the amounts and types of chemicals on siteso that DHS can determine whether a facility will be further regulated byCFATS rules. Regulated facilities are placed into one of four tiers, based on risk and performance. The facilities placed in the highest-risk tier will be subject to the most stringent security required. Pursuant to P.L. 109-295, the CFATS regulations will sunset in October 2009.

The 110th Congress

Because of the scope and time limitation of the Department's authority to regulate chemical facilities, as well as implementation issues, chemical facility security has remained on the congressional agenda. In particular, the interim final rule issued byDHS stated that it may preempt future state and local chemical facility security regulations. In response to this preemption language, the 110th Congress amended P.L. 109-295 in the Consolidated Appropriations Act, 2008 (P.L. 110-161) to establish a state's right to promulgate chemical facility security regulations that are at least as stringent as the federal chemical facility security regulations. P.L. 110-161 further provides that state regulations would be preempted only in the case of an actual conflict between federal and state regulations.

Several other bills have been introduced to modify DHS authority to regulate chemical facilities. Perhaps of most interest to the water sector is H.R. 5577, the Chemical Facility Anti-Terrorism Act of 2008. Reported, amended, from the House Committee on Homeland Security in March 2008, H.R. 5577 would revise, expand, and make permanent DHS's authority to regulate chemical facility security. Currently, water and wastewater facilities are explicitly exempt from P.L. 109-295 and the CFATS regulations; however, these utilities are not categorically exempt under H.R. 5577. This legislation would direct the Secretary to maintain a list of covered chemical facilities that are determined to be of sufficient security risk, based on several criteria: 1) the likelihood that the facility will be the target of an attack; 2) the potential extent and likelihood of death or injury, or harm to the environment, critical infrastructure, national security, or the economy that could result from an incident; and 3) the proximity of the chemical facility to population centers. DHS would be authorized to designate any chemical substance as a "substance of concern" and to establish and revise the threshold quantity for designated substances.

H.R. 5577 also would require owners and operators of covered facilities to conduct vulnerability assessments and to develop and implement DHS-approved security plans. To address the possibility of redundant federal requirements, DHS would be authorized to consider whether requirements under other laws (such as the Bioterrorism Preparedness Act) satisfy these requirements. Another provision would require owners and operators of affected facilities to complete an analysis of "methods to reduce the consequences of a terrorist attack" (the bill does not use the term "inherently safer technologies"), and would authorize the DHS Secretary to require higher risk facilities to implement such methods or technologies under certain conditions. As amended in markup, the Secretary would not be authorized to order affected drinking water or wastewater facilities to implement inherently safer technologies unless DHS provided funding to assist with the conversion. Also of relevance to the water sector is the provision authorizing DHS to order a facility assigned to the high-risk tier to cease operations if the facility fails to comply with a DHS compliance order. As reported, H.R. 5577 provides that DHS would be authorized to order the shut down of a water facility only if the facility's continued operation presented a clear danger to homeland security. Regulations promulgated under this bill would become effective when the current statutory authority expired in October 2009.[23]

The House Committee on Energy and Commerce, which traditionally has held jurisdiction over drinking water safety issues, was granted an extension for consideration of H.R. 5577 through May 2008. The bill has not been referred to the House Committee on Transportation and Infrastructure, which has key jurisdiction over wastewater utilities.

Assistance for Security Measures

Also at issue is the availability of funding for water systems to make security upgrades needed to address risks identified in their vulnerability assessments. Based on a limited assessment, EPA reported in 2005 that community water systems would need more than $1 billion to make security improvements.[24] The AWWA estimated that municipal water systems would have to spend more than $1.6 billion just to ensure control of access to critical water system assets.[25] Congress has not provided funding specifically for this purpose.[26] Although community water systems potentially are eligible to receive funding from the states through the DHS State Homeland Security Grant Program, competition for funds has been severe, and funds largely have gone to meet the needs of first responders. In an effort to address one element of this funding issue, the conference report for the Department of Homeland Security Appropriations Act for FY2005 (P.L. 108-334, H.Rept. 108-774) modified the definition of "local unit of government" specifically to include water districts.[27] Congress provided $525 million for this grant program for FY2007 (P.L. 109-295), and $950 million for FY2008 (P.L. 110-161).

In the 110th Congress, the Water SecurityAct of 2007, S. 1968, would authorize EPA to provide grants to drinking water and wastewater systems for security-related measures including vulnerability assessments, implementation of security enhancements, and developing or upgrading emergency response and site security plans. S. 1968 also would authorize EPA to provide technical assistance to small utilities, and would require research on specific security topics. A similar bill, S. 1303, targets wastewater treatment facilities. S. 1303 would authorize EPA to make grants to municipal wastewater treatment utilities for vulnerability assessments, security enhancements, emergency response plans, and facility security plans. The bill also would direct EPA to conduct research on wastewater infrastructure security. The chemical facility security bill, H.R. 5577, would authorize $225 million over three years to provide assistance to facilities that are required to adopt safer technologies to reduce the consequences of an attack.

The question of how to set priorities for allocating homeland security funding has been debated for several years. At a House Energy and Commerce Committee hearing on bioterrorism and the security of water supplies, the Government Accountability Office (GAO) testified that water security experts widely agreed that decisions for allocating federal funding for water security improvements should be based primarily on two criteria: 1) population density, and 2) information from vulnerability assessments.[28] Additionally, security experts established funding priorities at the utility level, identifying distribution systems as the most vulnerable component of a water system. Other water system components identified as requiring protective measures included utility computersystems, chemicals storedon-site, and source water supplies.[29]

Water Supply Security Research

A major security concern for the water sector has been the need for research to develop real-time monitoring methods to detect contaminants, and to develop technologies to remove or inactivate them. The water industry and security experts have identified a particular need for research to develop monitoring technologies that can quickly detect contaminants in water that has already left a treatment plant for distribution to consumers.

EPA has focused homeland security research on the detection, containment, and decontamination of chemical and biological agents that could be used in attacks on water systems. These efforts received limited support during the 109th Congress. For FY2006, EPA had requested $44 million for the Water Sentinel program (renamed theWater Security Initiative) to help address water utilities' concerns regardingtheir ability to detect and respond to chemical and biological contaminants.[30] The agency has considered this initiative to be a key element of its effort to meet its water security responsibilities under HSPD-9;[31] however, Congress provided just $8.1 million for the program for FY2006. In recommending a large reduction, the House Appropriations Committee commented that EPA should develop clear goals for the initiative, and justify the request more clearly in the budget request for FY2007.[32]

For FY2007, the Water Security Initiative (WSI) again failed to receive the requested support ($41.7 million) from House and Senate appropriators. Moreover, the House Appropriations Committee recommended that future funding for the initiative should be requested through DHS and not EPA. While agreeing that EPA's expertise was important for program success and that it had a critical role to play, the House committee expressed its strong belief that future program funding must be provided through the DHS.[33] Action was not completed on the EPA funding bill, and the continuing resolution for FY2007 (P.L. 110-5, H.J.Res. 20) generally funded EPA programs at the FY2006 level. However, EPA used its discretion and allocated $27 million to the Water Security Initiative, considerably more than the FY2006 appropriation. As discussed above in "Funding for Drinking Water Security Activities," the House and Senate appropriations committees again recommended reductions in the amount requested for this initiative for FY2008, but explained that the reductions were in response to the delayed obligation of FY2007 funds. Neither committee suggested that the initiative should be funded through the DHS, but both urged EPA to be prepared to report on the status and accomplishments of the initiative's pilot projects. The Consolidated Appropriations Act, 2008 (P.L. 110161), included $11.7 million for the initiative.

EPA has requested $21.4 million for the Water Security Initiative for FY2009, and proposes to use the resources to support two new WSI surveillance and monitoring pilots, bringing the total number of pilot projects to five. In its budget justification, the agency explains that five projects would be sufficient to generate the range of data needed for the program to have broad application to a variety of utilities in high-risk cities nationwide.[34] EPA also proposes to continue supporting research evaluating analytical methods for high risk contaminants and developing a network of laboratories to help confirm and speed responses to intentional contamination events; theagencyestimates that 90% of contaminants of high concern lack validated detection methods.

End Notes

[1] Another 19,174 public water systems are non-transient, non-community water systems (NTNCWS), such as schools or factories, that have their own water supply and generally serve the same individuals for more than six months but not year-round. More than 86,000 other public water systems are transient non-community water systems (TNCWS), such as campgrounds and gas stations, that provide their own water to transitory customers.

[2] For a broader review of security issues in the water resources sector (including dams and sewage treatment plants), see CRS Report RL32189, *Terrorism and Security Issues Facing the Water Infrastructure Sector*, by Claudia Copeland.

[3] For a review of critical infrastructures, related security issues and protection initiatives, and activities within the Department of Homeland Security, see CRS Report RL30153, *Critical Infrastructures: Background, Policy, and Implementation*, by John D. Moteff.

[4] The President's Commission on Critical Infrastructure Protection, *Critical Foundations: Protecting America's Infrastructures. Report of the President's Commission on Critical Infrastructure Protection*, Appendix A, Sector Summary Reports, October 1997, p. A-45.

[5] For further information on the Water ISAC, see [http://www.waterisac.org].

[6] EPA generally has not performed security training; rather, the agency has delivered training at locations across the country through stakeholder organizations and other federal partners. Organizations that have provided training include professional associations, such as the American Water Works Association, the Water Environment Federation, and the National Rural Water Association. Congress has provided some grant funds to these organizations, through EPA, to support their water security training activities.

[7] Water security reports, guidance documents, directives, and information on EPA security activities are available at [http://cfpub.epa.gov/safewater/watersecurity/index.cfm].

[8] U.S. Environmental Protection Agency, *Response Protocol Toolbox: Planning for and Responding to Drinking Water Contamination Threats and Incidents*, August 2004. Available at [http://www.epa.gov/watersecurity/pubs/rptb_response_guidelines.pdf].

[9] Office of Homeland Security, *The National Strategy for the Physical Protection of Critical Infrastructures and Key Assets*, February 2003, p. 39. The categories included (1) physical damage or destruction of critical assets (including the intentional release of toxic chemicals), (2) actual or threatened contamination of the water supply, (3) cyber attack, and (4) interruption of services from another infrastructure (such as energy supply).

[10] U.S. Environmental Protection Agency, *EPA's Role In Water Security Research: The Water Security Research and Technical Support Action Plan*, EPA/600/R-04/063, March 2004.

[11] National Academy of Sciences, *A Review of the EPA Water Security Research and Technical Support Action Plan: Parts I and II*, National Academy Press, 2003.

[12] U.S. Environmental Protection Agency, *Water Security Research and Technical Support Action Plan — Progress Report for 2005*, p. 5-7.

[13] U.S. Environmental Protection Agency, *FY2006 Annual Performance Plan and Congressional Justification*, Science and Technology, Homeland Security: Critical Infrastructure Protection, pp. S&T-21 - S&T-23.

[14] U.S. Congress, House Committee on Appropriations, *Department of the Interior, Environment, and Related Agencies Appropriation Bill, 2006*, report to accompany H.R. 2361, 109th Cong., 1st sess., H.Rept. 109-80, p. 94.

[15] U.S. Environmental Protection Agency, *FY2009 Annual Performance Plan and Congressional Justification*, Science and Technology, Homeland Security: Critical Infrastructure Protection, p. 70.

[16] See EPA Fact Sheet, *Use of the Drinking Water State Revolving Fund (DWSRF) to Implement Security Measures at Public Water Systems*, EPA 816-F-02-040, November 2001. Available at [http://www.epa.gov/safewater/dwsrf/pdfs/security-fs.pdf].

[17] For information on DHS grant programs, see CRS Report RL32348, *Selected Federal Homeland Security Assistance Programs: A Summary*, by Shawn Reese; and CRS Report RL34181, *Distribution of Homeland Security Grants in FY2007 and P.L. 110-53, Implementing Recommendations of the 9/11 Commission Act*, by Steven Maguire and Shawn Reese.

[18] EPA published *Water Security Strategy for Systems Serving Populations Less than 100,000/15MGD or Less* (July 2002).

[19] For a review of DHS reorganization actions related to critical infrastructure protection, see CRS Report RL30153, *Critical Infrastructures: Background, Policy, and Implementation*, by John D. Moteff.

[20] The NIPP can be found at [http://www.dhs.gov/xprevprot/programs/editorial_0827.shtm].

[21] The Water Sector-Specific Plan is available online at [http://www.epa.gov/safewater/watersecurity/pubs/plan_security_watersectorspecificplan.pdf].

[22] SDWA §1433(b) states that emergency response plans "shall include, but not be limited to, plans, procedures, and identification of equipment that can be implemented or utilized in the event of a terrorist or other intentional attack on the public water system. The emergency response plan shall also include actions, procedures, and identification of equipment which can obviate or significantly lessen the impact of terrorist attacks or other intentional actions on the public health and the safetyand supplyof drinking water provided to communities and individuals." (42 U.S.C. 300i-2)

[23] For more information on this and other bills, see CRS Report RL33847, *Chemical Facility Security: Regulation and Issues for Congress*, by Dana A. Shea and Todd B. Tatelman.

[24] U.S. Environmental Protection Agency, *Drinking Water Infrastructure Needs Survey and Assessment: Third Report to Congress*, June 2005. EPA 816-R-05-001. EPA noted that many water systems had not adequately captured security needs when this assessment was conducted in 2003. The agency anticipates that security needs will be reported more completely in the next assessment.

[25] American Water Works Association, *Protecting Our Water: Drinking Water Security in American After 9/11*, Executive summary, 2003.

[26] As discussed on page 9 above, some security projects are eligible for funding under the EPA-administered drinking water SRF program.

[27] For information on DHS grant program funding for FY2008, see CRS Report RS22596, *FY2008 Appropriations for State and Local Homeland Security*, by Shawn Reese.

[28] U.S. Government Accountability Office, *Drinking Water: Experts' Views on How Future Federal Funding Can Best Be Spent to Improve Security*, Testimony before the Subcommittee on Environment and Hazardous Materials, Committee on Energy and Commerce, House of Representatives. September 30, 2004. Testimony was based on report of same title GAO-04-29, October 2003.

[29] Ibid. pp. 2-3.

[30] U.S. Environmental Protection Agency, *FY2006 Annual Performance Plan and Congressional Justification*, Science and Technology, Homeland Security: Critical Infrastructure Protection, p. S&T-21 - S&T-23.

[31] Homeland Security Presidential Directive-9 directs EPA to develop a comprehensive and fully coordinated surveillance and monitoring system for drinking water, including a laboratory network to support surveillance and response activities.

[32] U.S. Congress, House Committee on Appropriations, *Department of the Interior, Environment, and Related Agencies Appropriation Bill, 2006*, report to accompany H.R. 2361, 109th Cong., 1st sess., H.Rept. 109-80, p. 94.

[33] U.S. Congress, House Committee on Appropriations, *Department of the Interior, Environment, and Related Agencies Appropriation Bill, 2007*, report to accompany H.R. 5386, 109th Cong., 2nd sess., H.Rept. 109-465, p. 100-101.

[34] U.S. Environmental Protection Agency, *FY2009 Annual Performance Plan and Congressional Justification*, Science and Technology, Homeland Security: Critical Infrastructure Protection, p. 68-69.

In: Safeguarding the Nation's Drinking Water ISBN: 978-1-60741-234-2
Editors: Juan A. Schrock

Chapter 3

SAFE DRINKING WATER ACT (SDWA): A SUMMARY OF THE ACT AND ITS MAJOR REQUIREMENTS*

Mary Tiemann

SUMMARY

This chapter summarizes the Safe Drinking Water Act (SDWA) and its major programs and regulatory requirements. It excerpts, with several additions, the SDWA chapter of CRS Report RL30798, *Environmental Laws: Summaries of Major Statues Administered by the Environmental Protection Agency (EPA)*, which provides summaries of the principal environmental statutes administered by the Environmental Protection Agency (EPA). This chapter includes the drinking water security provisions added to the SDWA by the Public Heath Security and Bioterrorism Preparedness and Response Act of 2002 (P.L. 107-188).

The Safe Drinking Water Act, Title XIV of the Public Health Service Act, is the key federal law for protecting public water supplies from harmful contaminants. First enacted in 1974 and substantially amended in 1986 and 1996, the act is administered through programs that establish standards and

---

* This is an edited, reformatted and augmented version of a CRS Report for Congress, Report RL31243, dated November 20, 2008.

treatment requirements for public water supplies, control underground injection of wastes, finance infrastructure projects, and protect sources of drinking water. The 1974 law established the current federal-state arrangement in which states may be delegated primary implementation and enforcement authority for the drinking water program. The state-administered Public Water Supply Supervision Program remains the basic program for regulating the nation's public water systems, and 49 states have assumed this authority.

The last major reauthorization of the act was done through the Safe Drinking Water Act Amendments of 1996 (P.L. 104-182), which generally authorized appropriations for SDWA programs through FY2003. As with other EPAadministered statutes having expired funding authority, Congress has.continued to appropriate funds for the ongoing SDWA programs.

In addition to reviewing key programs and requirements of the SDWA, this chapter includes statistics on the number and types of regulated public water systems. It also provides tables that list all major amendments, with the year of enactment and public law number, and that cross-reference sections of the act with the major U.S. Code sections of the codified statute.

INTRODUCTION

The Safe Drinking Water Act (SDWA), Title XIV of the Public Health Service Act, is the key federal law for protecting public water supplies from harmful contaminants. First enacted in 1974 and substantially amended in 1986 and 1996, the act is administered through programs that establish standards and treatment requirements for public water supplies, control underground injection of wastes, finance infrastructure projects, and protect sources of drinking water. The 1974 law established the current federal-state arrangement in which states may be delegated primary implementation and enforcement authority for the drinking water program. The state-administered Public Water Supply Supervision (PWSS) Program remains the basic program for regulating the nation's public water systems, and 49 states have assumed this authority. The 1996 SDWA amendments generally authorized appropriations for SDWA programs through FY2003. **Table 1** identifies the original enactment and subsequent amendments.

This chapter summarizes the act's major provisions, programs, and requirements, and is adapted from a broader document, CRS Report RL30798, *Environmental Protection Laws: Summaries of Statutes Administered by the Environmental Protection Agency*. It also provides selected statistics on the universe of regulated public water systems, and lists references for further information on the act and its implementation. **Table 5**, located at the end of this chapter, cites the major U.S. Code sections of the act and the equivalent sections of the statute.

BACKGROUND

As indicated by **Table 1**, the Safe Drinking Water Act has been amended several times since enactment of the Safe Drinking Water Act of 1974 (P.L. 93-523). Congress passed this law after nationwide studies of community water systems revealed widespread water quality problems and health risks resulting from poor operating procedures, inadequate facilities, and uneven management of public water supplies in communities of all sizes. The 1974 law gave the EPA substantial discretionary authority to regulate drinking water contaminants and gave states the lead role in implementation and enforcement.

The first major amendments (P.L. 99-339), enacted in 1986, were largely intended to increase the pace at which the EPA regulated contaminants and to increase the protection of ground water. From 1974 until 1986, the EPA had regulated just one additional contaminant beyond the 22 standards previously developed by the Public Health Service. The 1986 amendments required the EPA to (1) issue regulations for 83 specified contaminants by June 1989 and for 25 more contaminants every three years thereafter, (2) promulgate requirements for disinfection and filtration of public water supplies, (3) ban the use of lead pipes and lead solder in new drinking water systems, (4) establish an elective wellhead protection program around public wells, (5) establish a demonstration grant program for state and local authorities having designated sole-source aquifers to develop groundwater protection programs, and (6) issue rules for monitoring injection wells that inject wastes below a drinking water source. The amendments also increased EPA's enforcement authority.

Congress again amended SDWA with the Lead Contamination Control Act of 1988 (P.L. 100-572). These provisions were intended to reduce exposure to lead in drinking water by requiring the recall of lead-lined water coolers, and requiring the EPA to issue a guidance document and testing protocol for states to help schools and day care centers identify and correct lead contamination in school drinking water.

Table 1. Safe Drinking Water Act and Amendments (codified generally as 42 U.S.C. 300f-300j)

Year	Act	Public Law Number
1974	Safe Drinking Water Act of 1974	P.L. 93-523
1977	Safe Drinking Water Act Amendments of 1977	P.L. 95-190
1979	Safe Drinking Water Act Amendments	P.L. 96-63
1980	Safe Drinking Water Act Amendments	P.L. 96-502
1986	Safe Drinking Water Act Amendments of 1986	P.L. 99-339
1988	Lead Contamination Control Act of 1988	P.L. 100-572
1996	Safe Drinking Water Act Amendments of 1996	P.L. 104-182
2002	Public Health Security and Bioterrorism Preparedness and Response Act of 2002	P.L. 107-188

After the regulatory schedule mandated in the 1986 amendments proved to be unworkable for the EPA, states, and public water systems, the 104th Congress made sweeping changes to the act with the SDWA Amendments of 1996 (P.L. 104-182). As over-arching themes, the amendments targeted resources to address the greatest health risks, added some regulatory flexibility, provided funding for federal drinking water mandates, and aimed to improve water system compliance capacity. Congress revoked the requirement that the EPA regulate 25 new contaminants every three years and created a risk-based approach for selecting contaminants for regulation. Among other changes, Congress added some flexibility to the standard setting process, required the EPA to conduct health risk reduction and cost analyses for new rules, authorized a drinking water state revolving loan fund (DWSRF) program to help water systems finance projects needed to meet SDWA requirements, added programs to improve small system compliance, expanded consumer information requirements, increased the act's focus on pollution prevention through a voluntary source water protection program, and streamlined the act's enforcement provisions. P.L. 104-182 extended authorizations for appropriations under the act through FY2003.

In 2002, several drinking water security provisions were added to the SDWA through the Public Heath Security and Bioterrorism Preparedness and Response Act of 2002 (P.L. 107-188). Title IV of the act included requirements for community water systems serving more than 3,300 individuals to conduct vulnerability assessments and prepare emergency response plans. The law also required the EPA to conduct research on preventing and responding to terrorist or other attacks.

REGULATED PUBLIC WATER SYSTEMS

Federal drinking water regulations apply to the approximately158,000 privately and publicly owned water systems that provide piped water for human consumption to at least 15 service connections or that regularly serve at least 25 people. Nearly 53,000 of these systems are *community water systems* (CWSs) that serve the same residences year-round. These 53,000 systems provide water to approximately 282 million people. All federal regulations apply to these systems. Another 19,174 public water systems are *non-transient, non-community water systems* (NTNCWS), such as schools or factories, that have their own water supply and generally serve the same individuals for more than six months but not year-round. Most drinking water regulations apply to these systems. More than 86,200 other public water systems are *transient non-community water systems* (TNCWS), such as campgrounds and gas stations, that provide their own water to transitory customers. Only regulations for contaminants that pose immediate health risks apply to these systems.[1]

Table 2. Size Categories of Community Water Systems

System Size (population served)	Number of Community Water Systems	Percentage of Community Water Systems	Population Served (millions)	Percentage of Population Served
Very small (25-500)	29,666	56%	4.9	2%
Small (501-3,300)	14,389	27%	20.8	7%
Medium (3,301-10,000)	4,748	9%	27.5	10%
Large (10,001-100,000)	3,648	7%	102.7	36%
Very large (>100,000)	386	1%	126.3	45%
Total	54,064	100%	282.3	100%

Source: U.S. Environmental Protection Agency. EPA Safe Drinking Water Information System, *Factoids. Drinking Water and Ground Water Statistics for 2005.* EPA 816-K-03-001. Dec. 2006. Available on the Internet at [http://www.epa.gov/safewater/data/getdata.html].

Most community water systems (85%) are relatively small, serving 3,300 people or fewer; these systems provide water to just 9% of the total population served by community water systems. In contrast, 8% of systems serve populations of 10,000 or more, but provide water to 81% of the population served. Among the community water systems, 77% rely on ground water and 23% rely on surface water. (See **Table 2** for statistics on community water systems.)

More than 99% (19,055) of the non-transient, non-community water systems are small or very small and provide water to 83% of the population served by these systems. Approximately 99,700 of the nearly 105,400 non-community water systems (transient and non-transient systems combined) serve 500 or fewer people. These statistics give some insight into the scope of technological, economic, and managerial challenges small public water systems may face in meeting federal drinking water regulations. **Tables 3** and **4** provide statistics for non-transient non-community water systems and transient non-community water systems.

Table 3. Non-Transient Non-Community Water Systems (NTNCWS)

System Size (population served)	Number of NTNCWS	Percentage of NTNCWS	Population Served (millions)	Percentage of Population Served
Very small (25-500)	16,348	85%	2.28	38%
Small (501-3,300)	2,707	14%	2.71	45%
Medium (3,301-10,000)	102	1%	0.56	9%
Large (10,001-100,000)	17	0%	0.50	8%
Very large (>100,000)	0	0%	0	0%
Total	19,174	100%	6.92	100%

Source: EPA Safe Drinking Water Information System, *Factoids. Drinking Water and Ground Water Statistics for 2005*. NTNCWS regularly supplies water to at least 25 of the same people at least six months per year, but not year-round (e.g., schools, factories, office buildings and hospitals which have their own water systems. Most regulatory requirements apply to these systems.

Table 4. Transient Non-Community Water Systems (TNCWS)

System size (population served)	Number of TNCWS	Percentage of TNCWS	Population Served (millions)	Percentage of Population Served
Very small (25-500)	83,351	97%	7.29	52%
Small (501-3,300)	2,721	3%	2.67	19%
Medium (3,301-10,000)	111	0%	0.59	4%
Large (10,001-100,000)	23	0%	0.60	4 %
Very large (>100,000)	4	0%	2.99	21%
Total	86,210	100%	14.16	100%

Source: EPA Safe Drinking Water Information System, *Factoids. Drinking Water and Ground Water Statistics for 2005.* Transient non-community water systems provide water in places where people do not remain for long periods of time, such as gas stations and campgrounds. Only regulations for contaminants that pose immediate health risks (e.g. nitrate and bacteria) apply to these systems.

National Drinking Water Regulations

A key component of SDWA is the requirement that the EPA promulgate national primary drinking water regulations for contaminants that may pose health risks and that are likely to be present in public water supplies. Section 1412 instructs the EPA on how to select contaminants for regulation and specifies how the EPA must establish regulations once a contaminant has been selected. The regulations apply to the roughly 168,000 privately and publicly owned water systems that provide piped water for human consumption to at least 15 service connections or that regularly serve at least 25 people. The EPA has issued regulations for more than 90 contaminants, including regulations setting new standards for drinking water disinfectants and their byproducts and for microbial contaminants, a regulation establishing a standard for uranium in drinking water, and a regulation revising the standard for arsenic.

Contaminant Selection and Regulatory Schedules

Section 1412, as amended in 1996, directs the EPA to select contaminants for regulatory consideration based on occurrence, health effects, and meaningful opportunity for health risk reduction. Starting in 1998 and then every five years, the EPA must publish a list of contaminants that may warrant regulation. As of 2001, and every five years thereafter, the EPA must determine whether or not to regulate at least 5 of the listed contaminants. The act requires the EPA to evaluate contaminants that present the greatest health concern and to regulate those contaminants that occur at concentration levels and frequencies of public health concern. The amendments also included schedules for the EPA to complete regulations for specific contaminants (i.e., radon, arsenic, disinfectants and disinfection byproducts and *Cryptosporidium*).

Standard Setting

For each contaminant that the EPA determines requires regulation, the EPA must set a nonenforceable maximum contaminant level goal (MCLG) at a level at which no known or anticipated adverse health effects occur and which allows an adequate margin of safety. The EPA must then set an enforceable standard, a maximum contaminant level (MCL), as close to the MCLG as is "feasible" using best technology, treatment techniques, or other means available (taking costs into consideration). The agency generally sets standards based on technologies that are affordable for large communities; however, as amended by P.L. 104-182, the act requires EPA, when issuing a regulation for a contaminant, to list any technologies or other means that comply with the MCL and that are affordable for small public water systems serving populations of 10,000 or fewer. If the EPA does not identify "compliance" technologies that are affordable for small systems, then it must identify small system "variance" technologies or other means that may not achieve the MCL but are protective of public health.

The 1996 amendments authorized the EPA to set a standard at other than the feasible level if the feasible level would lead to an increase in health risks by increasing the concentration of other contaminants or by interfering with the treatment processes used to comply with other SDWA regulations. In such cases, the standard or treatment techniques must minimize the overall health risk. Also, when proposing a regulation, the EPA now must publish a determination as to whether or not the benefits of the standard justify the costs. If the EPA determines that the benefits do not justify the costs, the agency may, with certain exceptions, promulgate a standard that maximizes health risk reduction benefits at a cost that is justified by the benefits.

New regulations generally become effective three years after promulgation. Up to 2 additional years may be allowed if the EPA (or a state in the case of an individual system) determines the time is needed for capital improvements. (Section 1448 outlines procedures for judicial review of EPA actions involving the establishment of SDWA regulations and other final EPA actions.)

Risk Assessment

The 1996 amendments also added risk assessment and risk communication provisions to SDWA. When developing regulations, the EPA is required to: (1) use the best available, peer-reviewed science and supporting studies and data; and (2) make publicly available a risk assessment document that discusses estimated risks, uncertainties, and studies used in the assessment. When proposing drinking water regulations, the EPA must publish a health risk reduction and cost analysis (HRRCA). The EPA may promulgate an interim standard without first preparing this benefit-cost analysis or making a determination as to whether the benefits of a regulation would justify the costs if the Administrator determines that a contaminant presents an urgent threat to public health.

Variances and Exemptions

In anticipation that some systems, particularly smaller ones, could have difficulty complying with every regulation, Congress included in SDWA provisions for variances and exemptions. Section 1415 authorizes a state to grant a public water system a *variance* from a standard if raw water quality prevents meeting the standard despite application of best technology, and the variance does not result in an unreasonable risk to health. A 1996 provision (Subsection 1415(e)) authorizes variances specifically for small systems based on application of best affordable technology. When developing a regulation, if the EPA cannot identify a technology that meets the standard and is affordable for small systems, the EPA must identify variance technologies that are affordable but do not necessarily meet the standard. In cases where the EPA has identified variance technologies, then states may grant small system variances to systems serving 3,300 or fewer persons if the system cannot afford to comply with a standard (through treatment, an alternative water source, or restructuring) and the variance ensures adequate protection of public health. A state may then grant a variance to a small system, allowing the system to use a variance technology to comply with a regulation. With EPA approval, states also may grant variances to systems serving between 3,301 and 10,000 persons. Variances are not available for microbial contaminants. The EPA has determined that affordable compliance technologies are available for all existing standards. Thus, small system variances are not available.

Section 1416 authorizes states to grant public water systems temporary *exemptions* from standards or treatment techniques if a system cannot comply for other compelling reasons (including costs). An exemption is intended to give a water system more time to comply with a regulation and can be issued only if it will not result in an unreasonable health risk. A qualified system may receive an exemption for up to three years beyond the compliance deadline. Systems serving 3,300 or fewer persons may receive a maximum of three additional two-year extensions, for a total exemption duration of nine years.

STATE PRIMACY

Section 1413 authorizes states to assume primary oversight and enforcement responsibility (primacy) for public water systems. To assume primacy, states must adopt regulations at least as stringent as national requirements, develop adequate procedures for enforcement, adopt authority for administrative penalties, maintain records, and develop a plan for providing safe drinking water under emergency circumstances. Currently, 55 of 57 states and territories have primacy authority for the public water system supervision (PWSS) program.

Under Section 1443, Congress authorized appropriations of $100 million annually for the EPA to make grants to states to administer the PWSS program. This section directs the EPA, in accordance with regulations, to allot the sums among the states "on the basis of population, geographical area, number of public water systems, and other relevant factors." States are authorized to use a portion of their drinking water state revolving fund grant (under Section 1452) to cover the costs of administering the PWSS program.

Table 5. U.S. Code Sections of the Safe Drinking Water Act (Title XIV of the Public Health Service Act) (42 U.S.C. 300f-300j-26)

42 U.S.C.	Section Title	Safe Drinking Water Act (as amended)
Sub chapter XII	**Safety of Public Drinking Water Systems**	
Part A	Definitions	
300f	Definitions	sec. 1401
Part B	**Public Water Systems**	
300g	Coverage	sec. 1411
300g-1	National drinking water regulations	sec. 1412
300g-2	State primary enforcement responsibility	sec. 1413
300g-3	Enforcement of drinking water regulations	sec. 1414
300g-4	Variances	sec. 1415
300g-5	Exemptions	sec. 1416
300g-6	Prohibitions on the use of lead pipes, solder, and flux	sec. 1417
300g-7	Monitoring of contaminants	sec. 1418
300g-8[a]	Operator certification	sec. 1419
300g-9[a]	Capacity development	sec. 1420
Part C	**Protection of Underground Sources of Drinking Water**	
300h	Regulations for state programs	sec. 1421
300h-1	State primary enforcement responsibility	sec. 1422
300h-2	Enforcement of program	sec. 1423
300h-3	Interim regulation of underground injections	sec. 1424
300h-4	Optional demonstration by states relating to oil and natural gas	sec. 1425
300h-5	Regulation of state programs	sec. 1426
300h-6[a]	Sole source aquifer demonstration program	sec. 1427
300h-7[a]	State programs to establish wellhead protection areas	sec. 1428
300h-8[a]	State ground water protection grants	sec. 1429
Part D	**Emergency Powers**	
300i	Emergency powers	sec. 1431
300i-1	Tampering with public water systems	sec. 1432
300i-2[a]	Terrorist and other intentional acts	sec. 1433
300i-3	Contaminant prevention, detection, and response	sec. 1434

Table 5. (Continued)

42 U.S.C.	Section Title	Safe Drinking Water Act (as amended)
300i-4[a]	Supply disruption prevention, detection and response	sec. 1435
Part E	**General Provisions**	
300j	Assurance of availability of adequate supplies of chemicals necessary for treatment of water	sec. 1441
300j-1[a]	Research, technical assistance, information	sec. 1442
300j-2[a]	Grants for state programs	sec. 1443
300j-3[a]	Special project grants and guaranteed loans	sec. 1444
300j-4[a]	Records and inspections	sec. 1445
300j-5	National Drinking Water Advisory Council	sec. 1446
300j-6	Federal agencies	sec. 1447
300j-7	Judicial reviews	sec. 1448
300j-8	Citizen civil actions	sec. 1449
300j-9	General provisions	sec. 1450
300j-11	Indian Tribes	sec. 1451
300j-12[a]	State revolving loan funds	sec. 1452
300j-13	Source water quality assessment	sec. 1453
300j-14[a]	Source water petition program	sec. 1454
300j-15	Water conservation plan	sec. 1455
300j-16[a]	Assistance to colonias	sec. 1456
300j-17	Estrogenic substances screening program	sec. 1457
300j-18[a]	Drinking water studies	sec. 1458
Part F	**Additional Requirements to Regulate the Safety of Drinking Water**	
300j-21	Definitions	sec. 1461
300j-22	Recall of drinking water coolers with lead-lined tanks	sec. 1462
300j-23	Drinking water coolers containing lead	sec. 1463
300j-24	Lead contamination in school drinking water	sec. 1464
300j-25[a]	Federal assistance for state programs	sec. 1465
300j-26	Certification of testing laboratories	[b]

Note: This table shows only the major code sections. For more detail and to determine when a section was added, the reader should consult the official printed version of the U.S. Code.

a. These sections include authorizations of appropriations.

b. This provision was added by the Lead Contamination Control Act (P.L. 100-572, sec. 4),which did not amend SDWA.

ENFORCEMENT, CONSUMER INFORMATION, AND CITIZEN SUITS

The Safe Drinking Water Act requires public water systems to monitor their water supplies to ensure compliance with drinking water standards and to report monitoring results to the states. States review monitoring data submitted by public water systems, and also conduct their own monitoring, to determine system compliance with drinking water regulations. The EPA monitors public water system compliance primarily by reviewing the violation data submitted by the states.

Section 1414 requires that, whenever the EPA finds that a public water system in a state with primary enforcement authority does not comply with regulations, the agency must notify the state and the system and provide assistance to bring the system into compliance. If the state fails to commence enforcement action within 30 days after the notification, the EPA is authorized to issue an administrative order or commence a civil action. In a nonprimacy state, the EPA must notify an elected local official (if any has jurisdiction over the water system) before commencing an enforcement action against the system.

The 1996 amendments strengthened enforcement authorities, streamlined the process for issuing federal administrative orders, increased administrative penalty amounts, made more sections of the act clearly subject to EPA enforcement, and required states (as a condition of primacy) to have administrative penalty authority. The amendments also provided that no enforcement action may be taken against a public water system that has a plan to consolidate with another system.

Consumer Information and Reports

Enforcement provisions also require public water systems to notify customers of violations of drinking water standards or other requirements, such as monitoring and reporting requirements. Systems must notify customers within 24 hours of any violations that have the potential to cause serious health effects. Additionally, community water systems must mail to all customers an annual "consumer confidence report" on contaminants detected in their drinking water. States are required to prepare annual reports on the compliance of public water systems and to make summaries available to the EPA and the public; the EPA must prepare annual national compliance reports (Section 1414(c)).

Citizen Suits

Section 1449 provides for citizens' civil actions. Citizen suits may be brought against any person or agency allegedly in violation of provisions of the act, or against the EPA Administrator for alleged failure to perform any action or duty which is not discretionary.

Compliance Improvement Programs

The 1996 amendments added two state-administered programs aimed at improving public water system compliance with drinking water regulations: the operator certification program and the capacity development program. Section 1419 required states to adopt programs for training and certifying operators of community and nontransient noncommunity systems (e.g., schools and workplaces that have their own wells). The EPA is required to withhold 20% of a state's annual DWSRF grant, unless the state has adopted and implements an operator certification program. Relatedly, Section 1420 required states to establish capacity development programs, also based on EPA guidance. Congress specified that the programs must include (1) legal authority to ensure that new systems have the technical, financial, and managerial capacity to meet SDWA requirements; and (2) a strategy to assist existing systems that are experiencing difficulties to come into compliance. The EPA is required to withhold a portion of SRF grants from states that do not have capacity development strategies. The agency has not had to withhold funds under either of these programs.

Ground Water Protection Programs

Most public water systems rely on ground water as a source of drinking water, and Part C of the act focuses on ground water protection.[2] Section 1421 authorized the establishment of state underground injection control (UIC) programs to protect underground sources of drinking water (USDWs). In

1977, the EPA issued mandated regulations that contained minimum requirements for state UIC programs to prevent underground injection that endangers drinking water sources, and that required states to prohibit any underground injection not authorized by state permit. The law specified that the regulations could not interfere with the underground injection of brine from oil and gas production or recovery of oil unless underground sources of drinking water would be affected.[3]

Section 1422 authorized affected states to submit plans to the EPA for implementing UIC programs and, if approved, to assume primary enforcement responsibility. If a state's plan has not been approved, or the state has chosen not to assume program responsibility, then the EPA must implement the program (Section 1423). For oil and gas injection operations only, states with UIC programs are delegated primary enforcement authority without meeting EPA regulations, provided states demonstrate that they have an effective program that prevents underground injection that endangers drinking water sources (Section 1425). To implement this program, EPA placed UIC wells in five classes based on similarity in the fluids injected, and construction, injection depth, design, and operating techniques, and issued regulations that establish performance criteria for each class.[4] EPA has delegated primacy for all classes of wells to 34 states; it shares implementation responsibility in six states, and implements the UIC program for all well classes in 10 states.

Subsection 1424(e) authorizes the EPA to make determinations, on EPA's initiative or upon petition, that an aquifer is the sole or principal drinking water source for an area. In areas that overlie a designated sole-source aquifer, no federal funding may be committed for projects that the EPA determines may contaminate such an aquifer. Any person may petition for sole source aquifer designation. Nationwide, the EPA had designated 73 sole source aquifers.

The act contains three additional state programs aimed specifically at protecting ground water. Added in 1986, Section 1427 established procedures for demonstration programs to develop, implement, and assess critical aquifer protection areas already designated by the Administrator as sole source aquifers. Section 1428, also added in 1986, established an elective state program for protecting wellhead areas around public water system wells. If a state established a wellhead protection program by 1989, and the EPA approved the state's program, then the EPA may award grants covering between 50% and 90% of the costs of implementing the program. In 1996, Congress added Section 1429, authorizing the EPA to make 50% grants to states to develop programs to ensure coordinated and comprehensive protection of ground water within the states. Appropriations for these programs were authorized through FY2003 as follows: $15 million per year for Section 1427, $30 million per year for Section 1428, and $15 million per year for Section 1429.

Source Water Assessment and Protection Programs

The 1996 amendments expanded the act's pollution prevention focus to embrace protection of surface water, as well as ground water. Section 1453 required the EPA to publish guidance for states to implement source water assessment programs that delineate boundaries of the areas from which systems receive water, and identify the origins of contaminants in those areas to determine systems' susceptibility to contamination. States with approved assessment programs may adopt alternative monitoring requirements for water systems, as provided for in Section 1418.

Section 1454 authorized a source water petition program based on voluntary partnerships between state and local governments. States may establish a program under which a community water system or local government may submit a petition to the state requesting assistance in developing a voluntary source water quality protection partnership to (1) reduce the presence of contaminants in drinking water, (2) receive financial or technical assistance, and (3) develop a long-term source water protection strategy. This section authorized $5 million each year for grants to states to support petition programs. Also, states may use up to 10% of their DWSRF grant to support various source water protection activities including the petition program.

State Revolving Funds

In 1996, Congress authorized a drinking water state revolving loan fund (DWSRF) program to help systems finance improvements needed to comply with SDWA regulations (Section 1452). The EPA is authorized to make grants to states to capitalize DWSRFs, which states then may use to make loans to public water systems. States must match 20% of the federal grant. Grants are allotted based on the results of needs surveys. Each state and the District of Columbia must receive at least 1% of the appropriated funds.

Drinking water SRFs may be used to provide loans for expenditures that the EPA has determined will facilitate compliance or significantly further the act's health protection objectives. States must make available 15% of their annual allotment for loan assistance to systems that serve 10,000 or fewer persons, to the extent that funds can be obligated for eligible projects. States may use up to 30% of their DWSRF grant to provide loan subsidies (including forgiveness of principal) to help economically disadvantaged communities. Also, states may use a portion of funds for technical assistance, source water protection and capacity development programs, and for operator certification.

The law authorized appropriations of $599 million for FY1994 and $1 billion per year for FY1995 through FY2003 for DWSRF capitalization grants. The EPA was directed to reserve, from annual DWSRF appropriations, 0.33% for financial assistance to several Trusts and Territories, $10 million for health effects research on drinking water contaminants, $2 million for the costs of monitoring for unregulated contaminants, and up to 2% for technical assistance. The EPA may use 1.5% of funds each year for making grants to Indian Tribes and Alaska Native villages.[5]

Drinking Water Security

The 107th Congress passed the Public Health Security and Bioterrorism Preparedness and Response Act of 2002 (P.L. 107-188) to address a wide range of security issues. Title IV of the Bioterrorism Act amended SDWA to address threats to drinking water security. Key provisions are summarized below.

Vulnerability Assessments

Under new SDWA Section 1433, each community water system serving more than 3,300 people was required to conduct an assessment of the system's vulnerability to terrorist attacks or other intentional acts intended to disrupt the provision of a safe and reliable drinking water supply. This provision established deadlines, based on system size, for community water systems to certify to the EPA that they had conducted a vulnerability assessment and to submit to the EPA a copy of the assessment. Section 1433 exempted the contents of the vulnerability assessments from disclosure under the Freedom of Information Act (except for information contained in the certification identifying the system and the date of the certification), and it provides for civil and criminal penalties for inappropriate disclosure of information by government officials.

Section 1433 further required each of these community water systems to prepare an emergency response plan incorporating the results of the vulnerability assessment. The EPA was directed to provide guidance to smaller systems on how to conduct vulnerability assessments, prepare emergency response plans, and address threats.

Section 1433(e) authorized the appropriation of $160 million for FY2002, and such sums as may be necessary for FY2003 through FY2005, to provide financial assistance to community water systems to conduct vulnerability assessments, to prepare response plans, and for expenses and contracts to address basic security enhancements and significant threats.

The Bioterrorism Act also added Sections 1434 and 1435 to SDWA, directing the EPA Administrator to review methods by which terrorists or others could disrupt the provision of safe water supplies. The provisions requires the EPA to reiew methods for preventing, detecting, and responding to such disruptions, and methods for providing alternative drinking water supplies if a water system was destroyed or impaired. Section 1435(e) authorized $15 million for FY2002, and such sums as may be necessary for FY2003 through FY2005 to carry out Sections 1434 and 1435.

Emergency Powers

Under Section 1431, the Administrator has emergency powers to issue orders and commence civil action if (1) a contaminant likely to enter a public drinking water supply system poses a substantial threat to public health, and (2) state or local officials have not taken adequate action. The Bioterrorism Act amended this section to specify that EPA's emergency powers include the authority to act when there is a threatened or potential terrorist attack or other intentional act to disrupt the provision of safe drinking water or to impact the safety of a community's drinking water supply.

Tampering with Public Water Systems

Section 1432 provides for civil and criminal penalties against any person who tampers, attempts to tamper, or makes a threat to tamper with a public water system. Amendments made by the Bioterrorism Act increased criminal and civil penalties for tampering, attempting to tamper, or making threats to tamper with public water supplies. The maximum prison sentence for tampering increased from 5 to 20 years. The maximum prison sentence for attempting to tamper, or making threats to tamper, increased from 3 to 10 years. The maximum fine that may be imposed for tampering increased from $50,000 to $1 million. The maximum fine for attempting to tamper, or threatening to tamper, increased from $20,000 to $100,000.

Emergency Assistance

SDWA Subsection 1442(b) authorizes the EPA to provide technical assistance and to make grants to states and public water systems to assist in responding to and alleviating emergency situations. The Bioterrorism Act amended Subsection 1442(d) to authorize appropriations for such emergency assistance of not more than $35 million for FY2002, and such sums as may be necessary for each fiscal year thereafter.

ADDITIONAL SDWA PROVISIONS

Lead-free Plumbing

Section 1417 prohibits the use of any pipe, solder, or flux used in the installation or repair of public water systems or plumbing in residential or nonresidential facilities providing drinking water that is not "lead free" (as defined in the act). This section also makes it unlawful to sell pipes, plumbing fittings or fixtures that are not lead free, or to sell solder or flux that is not lead free (unless it is properly labeled), with the exception of pipes used in manufacturing or industrial processing.[6] Section 1417(e), added in 1996, directed EPA to promulgate regulations setting health-based performance standards limiting the amount of lead that may leach from new plumbing fittings and fixtures, unless a voluntary standard was established within one year of enactment. A voluntary standard was established.

Research, Technical Assistance, and Training

Section 1442 authorizes the EPA to conduct research, studies, and demonstrations related to the causes, treatment, control, and prevention of diseases resulting from contaminants in water. The agency is directed to provide technical assistance to the states and municipalities in administering their public water system regulatory responsibilities. This section authorized \$15 million annually for technical assistance to small systems and Indian Tribes, and \$25 million for health effects research. (Title II of P.L. 104-182, the 1996 amendments, authorized additional appropriations for drinking water research, not to exceed \$26.6 million annually for FY1997 through FY2003.)

Demonstration Grants

The Administrator may make grants to develop and demonstrate new technologies for providing safe drinking water and to investigate health implications involved in the reclamation/reuse of waste waters (Section 1444).

Records, Inspections, and Monitoring

Section 1445 states that persons subject to requirements under SDWA must establish and maintain records, conduct water monitoring, and provide any information that the Administrator may require by regulation to carry out the requirements of the act. Section 1445(b) authorizes the Administrator or a representative, after notifying the state in writing, to enter and inspect the property of water suppliers or other persons subject to the act's requirements, to determine whether the person is in compliance with the act. Failure to comply with these provisions may result in civil penalties.

This section also directs the EPA to promulgate regulations establishing the criteria for a monitoring program for unregulated contaminants. Beginning in 1999 and every five years thereafter, the EPA must issue a list of not more than 30 unregulated contaminants to be monitored by public water systems. States are permitted to develop representative monitoring plan to assess the occurrence of unregulated contaminants in small systems, and the section authorized \$10 million to be appropriated for each of FY1999 through FY2003 to provide grants to cover the costs of monitoring for small systems. All monitoring results are to be included in a national drinking water occurrence data base created under the 1996 amendments.

National Drinking Water Advisory Council

The act established a National Drinking Water Advisory Council, composed of 15 members (with at least two representing rural systems), to advise, consult, and make recommendations to the Administrator on activities and policies derived from the act (Section 1446).

Federal Agencies

Any federal agency having jurisdiction over federally owned public water systems must comply with all federal, state and local drinking water requirements as well as any underground injection control programs. The act provides for waivers in the interest of national security (Section 1447).

Assistance to Colonias[7]

Added in 1996, Section 1456 authorized the EPA and other appropriate federal agencies to award grants to Arizona, California, New Mexico and Texas to provide assistance (not more than 50% of project costs) to *colonias* where the residents are subject to a significant health risk attributable to the lack of access to an adequate and affordable drinking water system. Congress authorized appropriations of $25 million for each of fiscal years 1997 through 1999.

Estrogenic Substances

Section 1547 authorized the EPA to use the estrogenic substances screening program created in the Food Quality Protection Act of 1996 (P.L. 104-170) to provide for testing of substances that may be found in drinking water if the Administrator determines that a substantial population may be exposed to such substances.

Drinking Water Studies

Section 1458 directed the EPA to conduct drinking water studies involving subpopulations at greater risk and biological mechanisms. EPA also was directed to conduct studies to support specific regulations, including those for disinfectants and disinfection byproducts and *Cryptosporidium*.

SELECTED P.L. 104-182 PROVISIONS NOT AMENDING SDWA

The 104th Congress included a variety of drinking water-related provisions in the 1996 SDWA Amendments that did not amend the Safe Drinking Water Act. Several of these provisions are described below.

Transfer of Funds

Section 302 authorized states to transfer as much as 33% of their annual drinking water state revolving fund grant to the Clean Water Act (CWA) SRF, or an equivalent amount from the CWA SRF to the DWSRF through FY2001. (In subsequent conference reports for EPA appropriations, Congress has authorized states to continue transferring funds.)

Grants to Alaska

Section 303 of the 1996 amendments authorized the EPA to make grants to the State of Alaska to pay 50% of the costs of improving sanitation for rural and Alaska Native villages. Grants are for construction of public water and wastewater systems, and for training and technical assistance programs. Appropriations were authorized at $15 million for each of fiscal years 1997 through 2000. (In P.L. 106-457, Congress reauthorized appropriations for these rural sanitation grants at a level of $40 million for each of fiscal years 2001 through 2005.)

Bottled Water

Section 305 revised section 410 of the Federal Food, Drug, and Cosmetic Act to require the Secretary of Health and Human Services to issue bottled drinking water standards for contaminants regulated under SDWA within 180 days after the EPA promulgates the new standards, unless the Secretary determines that a standard is not necessary.

Wastewater Assistance to Colonias

Section 307 authorized EPA to make grants to colonias for wastewater treatment works. Appropriations were authorized at $25 million for each of fiscal years 1997 through 1999.

Additional Infrastructure Funding

Section 401 authorized additional assistance, up to $50 million for each of fiscal years 1997 through 2003, for a grant program for infrastructure and watershed protection projects.

Additional Reading

U.S. Congressional Budget Office. *Trends in Public Spending on Transportation and Water Infrastructure, 1956 to 2004.* Pub. No. 2880. August 2007. 38 p.

National Research Council. *Health Implications of Perchlorate Ingestion.* Board on Environmental Studies and Toxicology. National Academies Press. January 2005. 177 p.

U.S. Environmental Protection Agency. Office of Water. *Drinking Water State Revolving Fund: Increasing Impact,* 2006 Annual Report. EPA 816-R-07-002, June 2007. 44 p.

U.S. Environmental Protection Agency. Office of Water. *Drinking Water Infrastructure Needs Survey and Assessment: Third Report to Congress.* EPA 816-R-05-001. June 2005. 82 p.

U.S. Environmental Protection Agency. Office of Enforcement and Compliance Assurance. *Providing Safe Drinking Water in America: 2002*

National Public Water Systems Compliance Report. EPA 305-R-04-001. December 2004. 96 p.

U.S. Environmental Protection Agency. Office of Water. *Small System Guide to Safe Drinking Water Act Regulations: The First STEP to Providing Safe Drinking Water.* EPA 816-R-03-017. September 2003. 35 p.

U.S. Environmental Protection Agency. Office of Water. *The Clean Water and Drinking Water Infrastructure Gap Analysis Report.* EPA 816-R-02-020. September 2002. 50 p.

End Notes

[1] EPA's long-standing policy is to exclude transient systems from drinking water regulations except for those contaminants, such as nitrate, that the EPA believes have the potential to cause immediate adverse human health effects resulting from short-term exposure. (Source: Environmental Protection Agency. National Primary Drinking Water Regulation on Lead and Copper, minor revisions. January 12, 2000 (65 *FR* 1950).)

[2] According to the EPA, of roughly 156,600 public water systems, 142,400 rely on ground water and 14,200 rely on surface water. Among roughly 51,700 community water systems, 40,000 rely on ground water and 11,700 rely on surface water.

[3] The Energy Policy Act of 2005 (P.L. 109-58, § 322) amended SDWA § 1421(d) to specify that the definition of "underground injection" excludes the injection of fluids or propping agents (other than diesel fuels) used in hydraulic fracturing operations related to oil, gas, or geothermal production activities.

[4] The wells are classified as follows: Class I (inject hazardous wastes, industrial non-hazardous liquids, or municipal wastewater beneath the lowermost USDW); Class II (inject brines and other fluids associated with oil and gas production, and hydrocarbons for storage); Class III (inject fluids associated with solution mining of minerals beneath the lowermost USDW); Class IV (inject hazardous or radioactive wastes into or above USDWs and generally are banned); and Class V (all other injection wells — many of these wells inject non-hazardous fluids into or above USDWs and are typically shallow, on-site disposal systems).

[5] For more information, see CRS Report RS22037, *Drinking Water State Revolving Fund: Program Overview and Issues*, by Mary Tiemann.

[6] §1417(d) defines "lead free" to mean not more than 0.2% lead for solders and fluxes, and not more than 8% lead for pipes and pipe fittings.

[7] Colonias generally are described as unincorporated communities or housing developments on the U.S. side of the U.S.-Mexico border that lack some or all basic infrastructure including plumbing and public water and sewer.

In: Safeguarding the Nation's Drinking Water ISBN: 978-1-60741-234-2
Editors: Juan A. Schrock

Chapter 4

DRINKING WATER STATE REVOLVING FUND (DWSRF): PROGRAM OVERVIEW AND ISSUES*

Mary Tiemann

SUMMARY

The Safe Drinking Water Act (SDWA) Amendments of 1996 authorized a drinking water state revolving loan fund (DWSRF) program to help public water systems finance infrastructure projects needed to comply with federal drinking water regulations and to protect public health. Under the program, states receive capitalization grants to make loans to water systems for drinking water projects and certain other SDWA activities. Since FY1997, Congress has provided more than $10.3 billion for this program, including $829 million for FY2008. As of June 2007, the DWSRF program had provided a total of $12.6 billion in assistance and supported 5,346 projects.

The Environmental Protection Agency's (EPA's) latest (2003) survey of capital improvement needs for public water systems found that water systems need to invest $276.8 billion on infrastructure improvements over 20 years to ensure the provision of safe water. This amount represents a 60% increase over the 1999 survey estimate of $165.5 billion (in 2003 dollars) and reflects

---

* This is an edited, reformatted and augmented version of a CRS Report for Congress, Report RS22037, dated November 25, 2008.

funds needed for compliance with several revised regulations (e.g., the arsenic and radium rules), security-related needs, and better reporting of needs for infrastructure rehabilitation and replacement. Key issues include the gap between estimated needs and funding, SDWA compliance costs, and the need for cities to update and maintain water infrastructure, apart from SDWA compliance.

Congress substantially revised the Safe Drinking Water Act with the 1996 SDWA amendments (P.L. 104-182). A key provision, Section 1452, authorized a drinking water state revolving loan fund (DWSRF) program to help public water systems finance improvements needed to comply with federal drinking water regulations and to address the most serious risks to human health. The law authorized EPA to make grants to states to capitalize DWSRFs. States must match 20% of their annual grant and develop intended use plans each year indicating how the allotted funds will be used. States may use the DWSRF to provide loans and other assistance to eligible public water systems for expenditures that EPA has determined will facilitate SDWA compliance or significantly further the act's health protection objectives. Eligible projects include installation and replacement of failing treatment facilities, distribution systems, and certain storage facilities. Projects to replace aging infrastructure are eligible if they are needed to maintain compliance or to further public health protection goals. Projects to consolidate water supplies also may be eligible. This program is patterned after the 1987 Clean Water Act SRF (CWSRF) program for financing municipal wastewater treatment projects.

Public water systems eligible to receive DWSRF assistance include community water systems (whether publicly or privately owned) and not-for-profit noncommunity water systems.[1] States generally may not provide DWSRF assistance to systems that lack the capacity to ensure compliance with the act or that are in significant noncompliance with SDWA requirements, unless these systems meet certain conditions to return to compliance. Systems owned by federal agencies are not eligible. Although the law authorizes assistance to privately owned community water systems, some states have laws or policies that preclude privately owned utilities from receiving DWSRF assistance.

The 1996 amendments authorized appropriations for the DWSRF program of $599 million for FY1994 and $1 billion for each of FY1995 through FY2003. For each of FY2006 and FY2007, Congress funded the program at $837.5 million.[2] For FY2008, the President requested $842.2 million. The Consolidated Appropriations Act, 2008 (P.L. 110-161), included $829.0

million ($842.2 million before applying a 1.56% rescission). For FY2009, the President requested $842 million. The program is being funded under a continuing resolution (P.L. 110-329) at FY2008 levels through March 6, 2009.

Through June 2007, the EPA had awarded $8.13 billion in capitalization grants, which, when combined with the 20% state match, bond proceeds, loan principal repayments, and other funds, amounted to $13.9 billion in DWSRF funds available for loans and other assistance. Also as of June 2007, 5,346 projects had received assistance, and total assistance provided by the program reached $12.63 billion.[3]

DWSRF Allotments and Set-Asides

EPA is required to allot DWSRF funds among the states based on the results of the most recent quadrennial needs survey (discussed below). Each state and the District of Columbia must receive at least 1% of available funds, and as much as 0.33% of the total appropriation must be made available for grants to the Virgin Islands, the Commonwealth of the Northern Mariana Islands, American Samoa, and Guam. Before distributing funds among the states, EPA sets aside from the annual DWSRF appropriation $2 million to pay for monitoring of unregulated contaminants in small and medium systems, and 1.5% for grants to Indian Tribes and Alaska Native Villages ($12.56 million for FY2007). EPA is also authorized to reserve annually up to $30 million to reimburse states for operator training and certification costs if separate funding is not provided under Section 1419 of the SDWA; EPA reserved the full amount for several years, but reserved none after FY2003, as state training programs had matured. To provide technical assistance to small systems, EPA may reserve up to 2%, with a $15 million cap; however, funding for this activity is provided under Section 1442, and EPA has not set aside funds for this purpose.[4]

The law also includes several set-asides and directives that apply to states. These provisions offer states flexibility in tailoring their individual DWSRF programs to address state priorities. They also demonstrate the emphasis that the 1996 amendments placed on enhancing compliance, especially among smaller systems. The act requires states to make available at least 15% of their annual allotment for loan assistance to systems that serve 10,000 or fewer persons, to the extent that the funds can be obligated to eligible projects. The act also allows states to use up to 30% of their DWSRF grant to provide additional assistance, such as forgiveness of loan principal or negative interest rate loans, to help economically disadvantaged communities (as determined by the state).

Among other optional set-aside provisions, states may reserve as much as 4% of their DWSRF allotment to cover the costs of administering the DWSRF program and an additional portion to help pay the costs of other mandates added by the 1996 law. Specifically, states may set aside as much as 10% for a combination of the following: public water system supervision programs, technical assistance through source water protection programs, state capacity development strategies, and operator certification programs. To use DWSRF funds for these purposes, states must match these expenditures with an equal amount of state funds. States may use an additional 2% of funds to provide technical assistance to systems that serve 10,000 or fewer persons. States also have the option of using as much as 15% for a combination of the following: loans for the acquisition of land or conservation easements; loans to implement voluntary source water protection measures; technical and financial assistance to systems as part of a capacity development strategy; and development and implementation of ground water protection programs. Expenditures may not exceed 10% for any one of these activities. (Other SDWA provisions include funding authority for several of these programs and activities.)

To further enhance public water system compliance, the 1996 amendments added new capacity development and operator certification requirements. The law required EPA to withhold part of the DWSRF grant from any state that did not meet these mandates. Section 1420 required states to establish capacity development programs that include (1) legal authority or other means to ensure that new systems have the technical, financial, and managerial capacity to meet SDWA requirements and (2) a strategy to assist existing systems that are experiencing difficulties in coming into compliance. States also were required to adopt programs for training and certifying operators of community and nontransient non-community water systems.

Congress designed the DWSRF program to give states implementation flexibility. Congress gave states flexibility to set priorities between the SDWA and Clean Water Act SRF programs to accommodate the divergent drinking water and wastewater needs and priorities among the states. The law authorized states to transfer as much as 33% of the annual DWSRF allotment to the CWSRF or an equivalent amount from the CWSRF to the DWSRF. The statute authorized these transfers through FY2001. In 2000, EPA recommended that Congress continue to authorize transfers between the SRF programs to give states flexibility to address their most pressing water infrastructure needs. Subsequent conference reports for EPA appropriations have authorized states to continue transferring funds between these programs.

DRINKING WATER INFRASTRUCTURE NEEDS

SDWA section 1452(h) requires EPA to assess the capital improvement needs of eligible public water systems and to report to Congress every four years. Concurrently and in consultation with the Indian Health Service and Indian tribes, EPA must assess needs for drinking water treatment facilities to serve Indian tribes (section 1452(I)). EPA is required to distribute the DWSRF funds to the states based on the results of the latest needs survey. Eligible systems include approximately 53,000 community water systems and 21,400 not-for-profit noncommunity water systems.

EPA conducted its third survey of capital improvement needs for public water systems in 2003.[5] Based on this survey, EPA estimates that systems need to invest $276.8 billion on drinking water infrastructure improvements over 20 years to comply with drinking water regulations and/or to ensure the provision of safe water. This amount exceeds the 2001 needs survey estimate of $165.5 billion (in 2003 dollars) by more than 60%. The 2003 survey includes funds needed for compliance with several recent regulations (including the revised arsenic and radium rules) and pending rules for radon and other contaminants. It also identified $1 billion in security-related needs. Also, water systems made efforts to improve reporting of needs for infrastructure rehabilitation and replacement, which EPA determined had been under-reported in the previous surveys.

Of the total national need of $276.8 billion, $160.5 billion (60%) is currently needed to ensure the provision of safe drinking water. EPA notes that a "current need" typically involves installing, upgrading, or replacing infrastructure to allow a system to continue to deliver safe drinking water and that systems with current needs are usually not in violation of a drinking water standard. EPA reports that, although all of the infrastructure projects in the needs assessment promote the health objectives of the act, $45.1 billion (16%) of the total is attributable to SDWA regulations, while $237 billion (84%) represents nonregulatory costs (e.g., routine replacement of basic infrastructure).

The survey presents the 20-year needs estimates by category: transmission and distribution, treatment, source, storage, and other. The largest needs category, installation and rehabilitation of transmission and distribution systems, accounts for $183.6 billion (two-thirds) of total 20-year needs. Water treatment needs constituted the next largest category, accounting for $53.2 billion of total needs, while water storage accounts for $24.8 billion, and source (projects needed to obtain safe water supplies, including rehabilitation and installation of wells) accounts for $12.8 billion of total 20-year needs.

For further perspective, the needs survey breaks down the 20-year needs estimates according to system size and ownership. Large systems (serving more than 50,000 people) account for $122.9 billion of total 20-year need; medium systems (serving from 3,301 to 50,000 people) account for $103.0 billion; and small systems (serving 3,300 or fewer people) account for $34.2 billion. Not-for-profit noncommunity water systems have estimated needs of $3.4 billion. American Indian and Alaska Native Village water systems have estimated 20-year needs of $1.3 billion and $1.2 billion, respectively.

EPA notes that the total needs estimate may be conservative for several reasons: (1) systems are required to meet stringent documentation criteria when identifying needs; (2) many systems did not fully understand their security needs at the time of the assessment; (3) capital improvement plans often cover fewer than 10 years, while the survey tries to capture 20-year estimates; and (4) the survey is limited to eligible needs, thus excluding projects related to dams, raw water reservoirs, fire protection, operation and maintenance, and future growth.

Other needs assessments have also been prepared, including EPA's 2002 Gap Analysis. This study identified potential funding gaps between projected needs and spending from 2000 through 2019. EPA estimated a potential 20-year funding gap for drinking water capital and operations and maintenance ranging from $45 billion to $263 billion, depending on different scenarios.[6] (For more information on this study and other needs assessments, see CRS Report RL31116, *Water Infrastructure Needs and Investment: Review and Analysis of Key Issues*, by Claudia Copeland and Mary Tiemann.)

PROGRAM ISSUES

With the authorization of the DWSRF program, Congress acted to help public water systems finance infrastructure projects needed to achieve or maintain compliance with SDWA requirements and protect public health. While this federal/state program provides an important means for addressing drinking water needs, a substantial gap remains between financing needs and available funds. The 2003 needs survey identified $276.8 billion in drinking water infrastructure needs over 20 years, while the DWSRF program was authorized at $9.6 billion over seven years. The appropriated amounts, augmented by the state match, leveraging, repayments, and interest earnings, have created significant financing capacity among the state DWSRFs. However, many expect a funding gap to persist, and new SDWA requirements are expected to drive up future estimates of needs.

Other SDWA mandates are eligible for DWSRF funding and heighten competition for these resources. The DWSRF program embraces competing objectives, and thus, this competition is perhaps unavoidable. On the one hand, the fundamental purpose of the program is to capitalize revolving funds in the states in order to generate a perpetual source of funding for drinking water projects. On the other hand, Congress authorized multiple set-asides to fund other drinking water program priorities and requirements, such as system compliance capacity assurance, operator certification, and small system technical assistance. Overall, states may use as much as 31% of their grant for the setasides and 30% to provide loan subsidies to economically disadvantaged communities. While these options offer states flexibility to tailor their programs to meet individual needs, using funds for these activities could significantly erode the corpus of state funds and slow the rate at which they become capitalized. A concern for states is that, to the degree that Congress relies on the DWSRF to fund other SDWA requirements instead of providing separate appropriations, the potential of the DWSRF program is diminished.

A separate issue is the need for communities to address drinking water infrastructure needs that are outside the scope of the DWSRF program. Communities typically must address several categories of infrastructure requirements unrelated to SDWA compliance and, thus, ineligible for DWSRF assistance. These categories include future growth, ongoing rehabilitation, and operation and maintenance of systems. EPA has reported that outdated and deteriorated drinking water infrastructure poses a fundamental long-term threat to drinking water safety, and that in many communities, basic infrastructure costs can far exceed SDWA compliance costs. Although the DWSRF program does not address certain categories of needs and excludes many noncommunity water systems from coverage, with this program Congress has added a major tool to the mix of federal, state, and local initiatives intended to help communities ensure the safety of water supplies.

Ongoing drinking water infrastructure issues include the gap between funding and estimated needs; the growing cost of complying with SDWA standards, particularly for small communities; the ability of small or economically disadvantaged communities to afford DWSRF financing; and the broader need for cities to maintain, upgrade, and expand infrastructure unrelated to SDWA compliance. Despite congressional interest in recent years, budgetary constraints have posed challenges to efforts to enact water infrastructure funding legislation. In the face of large needs, scarce federal resources, and debate over the federal role in funding water infrastructure, EPA, states, and utilities have increasingly focused on alternative management and financing strategies to address costs and promote greater financial self-reliance among water systems. Strategies include establishing public-private partnerships, improving asset management, and adopting fullcost pricing for water services. These approaches are improving the sustainability of water systems; however, they may be limited in their ability to fully meet needs, particularly among poorer communities and small water systems. Thus, interest in infrastructure financing options and expanded federal assistance continues.

In recent years, House and Senate committees have held hearings on the SRF programs, infrastructure needs, and funding issues. For four Congresses, the Senate Environment and Public Works Committee has reported a water infrastructure financing bill, including S. 3617 (S.Rept. 110-509) in the 110th Congress. Similar to the committee bill from the 109th Congress, S. 3617 would have authorized increased funding for drinking water and wastewater SRF programs (authorizing $15 billion over five years for the DWSRF), allowed new uses for the funds, and would have created a grant program for small or economically disadvantaged communities for critical water projects. Other reported drinking water infrastructure bills included S. 1933 (S.Rept. 110-475), to create a grant program for small systems, and S. 199 (S.Rept. 110-476), to increase the authorization of appropriations for water and wastewater grants for Alaska's rural and Native villages. None of the bills was enacted.

End Notes

[1] A community water system is a system that serves at least 15 service connections used by yearround residents or that regularly serves at least 25 year-round residents. Other public water systems are noncommunity water systems (e.g., schools and workplaces with their own wells).

[2] Congress has provided $1.275 billion for FY1997 (the first year for which DWSRF authority was in place), $725 million for FY1998, $775 million for FY1999, $816.9 million for FY2000, $823.2 million for FY2001, $850 million for FY2002, $850 million for FY2003 ($844.5 million after applying the mandatory across-the-board 0.65% reduction in P.L. 108-7), $850 million for FY2004 ($844.9 million when adjusted for the 0.59% reduction under P.L. 108-199), and $850 million for FY2005 ($843.2 million after the 0.8% reduction in P.L. 108-447). For FY2006, P.L. 109-54 included $850 million ($837.5 million after applying two rescissions).

[3] Program statistics are available at [http://www.epa.gov/safewater/dwsrf/dwnims.html]. For further discussion of the DWSRF program, see EPA Report to Congress, *The Drinking Water State Revolving Fund Program*, May 2003, [http://www.epa.gov/safewater/dwsrf.html#Facts].

[4] DWSRF state allotments and set asides are available at EPA website, [http://www.epa.gov/safewater/dwsrf/allotments/index.html].

[5] Environmental Protection Agency, *Drinking Water Infrastructure Needs Survey and Assessment: Third Report to Congress*, June 2005. EPA 816-R-05-001. Available online at [http://www.epa.gov/safewater/needs.html].

[6] U.S. Environmental Protection Agency, *Clean Water and Drinking Water Infrastructure Gap Analysis Report*, EPA 816-R-02-020, September 2002.

In: Safeguarding the Nation's Drinking Water ISBN: 978-1-60741-234-2
Editors: Juan A. Schrock

Chapter 5

PERCHLORATE CONTAMINATION OF DRINKING WATER: REGULATORY ISSUES AND LEGISLATIVE ACTIONS*

Mary Tiemann

SUMMARY

Perchlorate is the explosive component of solid rocket fuel, fireworks, road flares, and other products. Used heavily by the Department of Defense (DOD) and related industries, perchlorate also occurs naturally and is present in some organic fertilizer. This soluble, persistent compound has been detected in drinking water supplies, especially in California. It also has been found in milk and many foods. Because of this widespread occurrence, concern over the potential health risks of perchlorate exposure has increased, and some states, water utilities, and Members of Congress have urged the Environmental Protection Agency (EPA) to set a federal drinking water standard for this chemical. Regulatory issues involve the relative health benefits and costs of federal regulation, including environmental cleanup and water treatment costs, both of which are driven by federal and state standards. (California and Massachusetts have set standards.) EPA has spent years assessing perchlorate's health effects and occurrence (including its occurrence in food)

* This is an edited, reformatted and augmented version of a CRS Report for Congress, Report RS21961, dated November 24, 2008.

to determine whether a national standard is warranted. Interagency disagreements over the risks of perchlorate exposure led several federal agencies to ask the National Research Council (NRC) to evaluate perchlorate's health effects and EPA's risk analyses. In 2005, the NRC issued its report, and EPA adopted the NRC's recommended reference dose (i.e., the expected safe dose) for perchlorate exposure. Subsequent studies raised more concerns about potential effects of low-level exposures, particularly for infants in certain cases. In October 2008, EPA announced its preliminary determination not to regulate perchlorate in drinking water. This chapter reviews perchlorate contamination issues and related developments.

BACKGROUND

Ammonium perchlorate is the key ingredient in solid fuel for rockets and missiles; other perchlorate salts are used to manufacture products such as fireworks, air bags, and road flares. Uncertainty about the health effects of perchlorate exposure has slowed efforts to establish drinking water and environmental cleanup standards. However, because of perchlorate's persistence in water and ability to affect thyroid function, concern has escalated with the detection of perchlorate in water in at least 33 states. In the absence of a federal standard, states have begun to adopt their own measures. Massachusetts set a drinking water standard of 2 parts per billion (ppb, or micrograms per liter [μg/L]) in 2006, and California adopted a 6 ppb standard in October 2007. Several states have issued health goals or advisory levels ranging from 1 ppb in Maryland (advisory level) and New Mexico (drinking water screening level) to 51 ppb in Texas (industrial cleanup level).

Occurrence

Perchlorate has been used heavily by the DOD and its contractors, and perchlorate contamination has been found near weapons and rocket fuel manufacturing facilities and disposal sites, research facilities, and military bases. Fireworks, road flares, construction sites, and other manufacturing activities and facilities also have been sources of contamination. Moreover, perchlorate occurs naturally (in West Texas, for example), is present in organic fertilizer imported from Chile, and can occur as a breakdown product of other products. It has been detected in drinking water sources, primarily in the Southwest and in scattered locations across the country. Contamination has been found most often in ground water, including some large aquifers in California.

In 1999, EPA required public water systems to monitor for perchlorate under the Unregulated Contaminant Monitoring Rule (UCMR) to determine the frequency and levels at which it is present in public water supplies nationwide. The UCMR required monitoring by all systems serving more than 10,000 persons and by a sample of smaller systems. Of some 3,700 water systems tested, perchlorate was detected in 153 systems in 26 states and two commonwealths, including 58 systems in California. In 14 systems, perchlorate levels exceeded EPA's preliminary remediation goal of 24.5 ppb. California has required more comprehensive monitoring, and perchlorate has been detected at least twice in 241 active or standby sources of drinking water in that state since 2002. In 2005, EPA reported perchlorate contamination at 65 DOD facilities, 7 other federal facilities, and 37 private sites.[1] All sampling results combined (i.e., soil, public and private drinking water wells, groundwater monitoring wells, and surface water), the Government Accountability Office reported that perchlorate has been found at 395 sites.[2]

Monitoring also has been conducted to assess the presence of perchlorate in foods. In 2004, the Food and Drug Administration (FDA) tested 500 samples of foods, including vegetables, milk, and bottled water for perchlorate. Samples were taken in areas where water was thought to be contaminated. The FDA found perchlorate in roughly 90% of lettuce samples (average levels ranged from 11.9 ppb to 7.7 ppb for lettuces), and in 101 of 104 bottled milk samples (with an average level of 5.7 ppb). To assess the presence of perchlorate in a wider range of foods, the FDA began testing all samples in its Total Diet Study in 2005.[3] Perchlorate was detected in 625 of 1065 (50%) of samples, and in 211 of the 285 (74%) foods tested.[4] In most cases, perchlorate levels were in the low single digits; however, levels were higher in some foods (e.g., shrimp, tomatoes, spinach, and bacon). The study found that 2-year-olds have the highest total perchlorate intake per kilogram body weight per day, followed by infants (6 to 12 months of age) and children 6 to 10 years of age. The widespread detection of perchlorate in food is relevant to EPA's standard-setting efforts, because EPA considers non-water exposures when determining whether to establish a standard for a contaminant, and at what level to set a standard.

Health Effects

Perchlorate is not known to cause cancer. It is known to disrupt the uptake of iodine in the thyroid, and health effects associated with perchlorate exposure are expected to parallel those caused by iodine deficiency. Iodine deficiency decreases the production of thyroid hormones, which help regulate the body's metabolism and growth. A key concern is that impairment of thyroid function in pregnant women can affect fetuses and nursing infants and can result in delayed development and decreased learning capacity. Several human studies have indicated that thyroid changes occur in humans at significantly higher levels of perchlorate than the amounts typically observed in water supplies.[5] However, a 2006 study by the Centers for Disease Control and Prevention (CDC) of a representative sample of the U.S. population found that environmental exposures to perchlorate have an effect on thyroid hormone levels in women with iodine deficiency. (No effect was found in men.) Fully 36% of the 1,111 women in this study were found to be iodine deficient, and the median level of urinary perchlorate measured in the women was 2.9 ppb.[6]

EPA Assessment of Perchlorate for Regulation

Over the past decade, EPA has evaluated perchlorate to determine whether a federal drinking water standard is needed. Under the Safe Drinking Water Act (SDWA, §1412(b)(1)), EPA must regulate a contaminant if the Administrator determines that the contaminant (1) may have an adverse health effect, (2) occurs in public water systems at a frequency and level of public health concern, and (3), in the sole judgment of the Administrator, regulation of the contaminant presents a meaningful opportunity for reducing health risks. In 1997, when a sensitive detection method became available for perchlorate and detections increased, scientific information was limited. In 1998, EPA placed perchlorate on the list of contaminants that were candidates for regulation, but concluded that information was insufficient to determine whether perchlorate should be regulated under the SDWA. EPA listed perchlorate as a priority for further research on health effects and treatment technologies and for collecting occurrence data. In 1999, EPA required water systems to monitor for perchlorate under the Unregulated Contaminant Monitoring Rule to determine the frequency and levels at which it is present in public water supplies nationwide. In January 2007, EPA reported that it had collected sufficient occurrence data, and that further monitoring was not needed for the agency to make a regulatory determination (72 *Fed. Reg.* 367, January 4, 2007).

Perchlorate Risk Assessment

In 1992, and again in 1995, EPA issued draft reference doses (RfDs) for perchlorate exposure. An RfD is an estimate (with uncertainty spanning perhaps an order of magnitude) of a daily oral exposure that is not expected to cause any adverse, non-cancer health effects during a lifetime. In developing an RfD, EPA incorporates factors to account for sensitive subpopulations, study duration, interand intraspecies variability, and data gaps. The draft RfDs range of 0.0001 to 0.0005 milligrams per kilogram (mg/kg) body weight per day translated to a drinking water equivalent level of 4 ppb-18 ppb. EPA takes the RfD into account when setting a drinking water standard; it also considers costs, the capabilities of monitoring and treatment technologies, and other sources of perchlorate exposure, such as food.

EPA's 1999 draft risk characterization resulted in a human risk benchmark of 0.0009 mg/kg per day (with a 100-fold uncertainty factor), which converted to a drinking water equivalent level of 32 ppb. However, EPA determined that the available health effects and toxicity database was inadequate for risk assessment. In 1999, EPA issued an *Interim Assessment Guidance for Perchlorate*, which recommended that EPA risk managers use the earlier reference dose range and drinking water equivalent level (DWEL) of 4-18 ppb for perchlorate-related assessment activities at hazardous waste sites.

In 2002, EPA prepared a draft risk assessment that concluded that the potential human health risks of perchlorate exposures include effects on the developing nervous system and thyroid tumors, based on rat studies that observed benign tumors and adverse effects in fetal brain development. The document included a draft RfD of 0.00003 mg/kg per day, which translated to a drinking water equivalent level of 1 ppb. This document was controversial, both for its implications for cleanup costs and for science policy reasons. (For example, some peer reviewers expressed concern over EPA's risk assessment methodology and reliance on rat studies.) DOD, water suppliers, and other commentors expressed concern that the draft RfD could lead to unnecessarily stringent and costly cleanups of perchlorate releases at federal facilities and in water supplies. In 2002, a federal interagency perchlorate working group convened to discuss perchlorate risk assessment, research and regulatory issues, and related agency concerns. Working group members included DOD, EPA, the Department of Energy, the National Aeronautics and Space Administration, the Office of Science and Technology Policy, the Council on Environmental Quality, and the Office of Management and Budget.

NRC Perchlorate Study

To resolve some of the uncertainty and debate over perchlorate's health effects and the 2002 draft risk assessment, the interagency working group asked the National Research Council (NRC) to review the available science for perchlorate and EPA's draft assessment. The NRC was asked to comment and make recommendations. The NRC Committee to Assess the Health Implications of Perchlorate Ingestion issued its review in January 2005 and suggested several changes to EPA's draft risk assessment. The committee concluded that because of key differences between rats and humans, studies in rats were of limited use for quantitatively assessing human health risk associated with perchlorate exposure. Although the committee agreed that thyroid tumors found in a few rats were likely perchlorate treatment-related, it concluded that perchlorate exposure is unlikely to lead to thyroid tumors in humans. The committee noted that, unlike rats, humans have multiple mechanisms to compensate for iodide deficiency and thyroid disorders. Also, the NRC found flaws in the design and methods used in the rat studies. The committee concluded that the animal data selected by EPA should not be used as the basis of the risk assessment.

The committee also reviewed EPA's risk assessment model. It agreed that EPA's model for perchlorate toxicity represented a possible early sequence of events after exposure, but it did not think that the model accurately represented possible outcomes after changes in thyroid hormone production. Further, the committee disagreed with EPA's definition of a change in thyroid hormone level as an adverse effect. Rather, the NRC defined transient changes in serum thyroid hormone as biochemical events that might precede adverse effects, and identified hypothyroidism as the first adverse effect.

Because of research gaps regarding perchlorate's potential effects following changes in thyroid hormone production, the committee made the recommendation that EPA use a *nonadverse effect* (i.e., the inhibition of iodide uptake by the thyroid in humans) rather than an adverse effect as the basis for the risk assessment. The committee explained that "[i]nhibition of iodide uptake is a more reliable and valid measure, it has been unequivocally demonstrated in humans exposed to perchlorate, and it is the key event that precedes all thyroid-mediated effects of perchlorate exposure."[7] Based on the use of this point of departure, the reliance on human studies, and the use of an uncertainty factor of 10 (for intraspecies differences), the NRC's recommendations led to an RfD of 0.0007 mg/kg per day. The committee concluded that this RfD should protect the most sensitive population (i.e., the fetuses of pregnant women who might have hypothyroidism or iodide deficiency) and noted that the RfD was supported by clinical studies, occupational and environmental epidemiologic studies, and studies of long-term perchlorate administration to patients with hyperthyroidism. In addition, the NRC identified data gaps and research needs. The committee has received some criticism for the extent to which it relied on a small, short-term human study, and debate over perchlorate's health risks has continued.

EPA's Response

In 2005, EPA adopted the NRC recommended reference dose of 0.0007 mg/kg per day, which translates to a drinking water equivalent level of 24.5 ppb. The DWEL is the concentration of a contaminant in water that is expected to have no adverse effects; it is intended to include a margin of safety to protect the fetuses of pregnant women who might have a preexisting thyroid condition or insufficient iodide intake. EPA based the DWEL on the assumption that all exposure would come from drinking water. If EPA were to develop a drinking water standard for perchlorate, it would lower the DWEL to account for other sources of exposure, particularly food.

In January 2006, EPA's Superfund office issued guidance adopting the NRC reference dose and the DWEL of 24.5 ppb as the recommended value to be considered as the preliminary remediation goal (PRG) to guide perchlorate assessment and cleanup at Superfund sites. In March, EPA's Children's Health Protection Advisory Committee (CHPAC) wrote to the EPA Administrator that the PRG does not protect infants, who are highly susceptible to neurodevelopmental toxicity and may be more exposed than fetuses to perchlorate. The CHPAC noted that perchlorate is concentrated in breast milk and that nursing infants could receive daily doses greater than the RfD if the mother is exposed to 24.5 ppb perchlorate in tap water. The committee recommended that the Superfund office lower the PRG and that the Office of Water develop a standard for perchlorate and, in the interim, issue a drinking water health advisory that takes into account early life exposures.

In October 2008, EPA announced a preliminary determination not to regulate perchlorate, noting that less than 1% of water systems have perchlorate levels above the health reference level. EPA concluded that perchlorate failed to meet two of SDWA's regulatory criteria (i.e., that a contaminant occur frequently at levels of health concern, and that establishing a national drinking water standard would provide a "meaningful opportunity for health risk reduction"). In response, EPA's Science Advisory Board's (SAB's) Drinking Water Committee argued that, given perchlorate's occurrence and welldocumented toxicity, EPA must have a compelling basis to support a determination not to regulate. The SAB requested more time to review the new model EPA relied on, and to comment on the preliminary determination. If the final determination is not to regulate, EPA will develop a nonenforceable health advisory to guide state and other officials on health effects, monitoring and treatment technologies, and expected safe exposure levels.

Department of Defense Perchlorate Actions

DOD is responsible for some large releases of perchlorate contamination and has allotted significant resources to address this problem. DOD has spent more than $114 million on research activities regarding perchlorate treatment technologies, detection methods, toxicity studies, and substitutes. Additional funds have been spent on cleanup.

Although remediation has proceeded at some sites, cleanups typically are driven by drinking water standards or other established cleanup standards. With no federal standard, cleanup goals and responsibilities have been ambiguous outside of California and Massachusetts where standards have been set. In 2006, after EPA established a DWEL for perchlorate, DOD adopted a policy setting 24 ppb as the level of concern to be used in managing perchlorate releases, unless a more stringent federal or state standard exists. The policy applies broadly to DOD installations and former military lands, and directs the services to test for perchlorate when it is reasonably expected that a release has occurred. If perchlorate levels exceed 24 ppb, a site-specific risk assessment must be conducted; if the assessment indicates that the perchlorate could result in adverse health effects, then the site must be prioritized for risk management. DOD uses a relative risk site evaluation framework to help prioritize environmental restoration work and to allocate resources among sites. (For more information, see DOD website: [https://www.denix.osd.mil/portal/page/portal/denix/environment/MERIT/EC/ECAL].)

CONGRESSIONAL ACTIONS

Several perchlorate bills were considered, but none were enacted during the 110th Congress. Responding to EPA's 2007 decision not to require further monitoring for perchlorate as an unregulated contaminant, S. 24 was introduced to require community water systems to test for perchlorate and disclose its presence in annual consumer reports. S. 150 and H.R. 1747 would have required EPA to set a standard for perchlorate. The Senate Environment and Public Works Committee reported S. 24 (S.Rept. 110-483) and S. 150 (S.Rept. 110-484). Additionally, H.Con.Res. 347 expressed the sense of Congress that the CDC and FDA should take action to educate the public on the importance of adequate iodine intake. (Iodine is protective against perchlorate exposure.)

End Notes

[1] EPA, Federal Facilities Restoration and Reuse, *Known Perchlorate Releases in the U.S.*, March 25, 2005, at [http://www.epa.gov/fedfac/documents/perchlorate_links.htm#occurrences].

[2] U.S. Government Accountability Office, *Perchlorate: A System to Track Sampling and Cleanup Results is Needed*, GAO-05-462, May 2005, pp. 29-44.

[3] The FDA test results are available online at [http://www.cfsan.fda.gov/~dms/clo4dat2.html].

[4] Clarence W. Murray et al., "U.S. Food and Drug Administration's Total Diet Study: Dietary Intake of Perchlorate and Iodine," *Journal of Exposure Science and Environmental Epidemiology* (2008), 1-10.

[5] Michael A. Kelsh et al., "Primary Congenital Hypothyroidism, Newborn Thyroid Function, and Environmental Perchlorate Exposure Among Residents of a Southern California Community," *Journal of Occupational Environmental Medicine*, 2003, p. 1117.

[6] Benjamin C. Blount, James L. Pirkle, et al., "Urinary Perchlorate and Thyroid Hormone Levels in Adolescent and Adult Men and Women Living in the United States," Centers for Disease Control and Prevention, in *Environmental Health Perspectives*, December 2006, p. 1865.

[7] National Research Council, *Health Implications of Perchlorate Ingestion*, Committee to Assess the Health Implications of Perchlorate Ingestion, National Academy of Sciences, 2005, p. 9.

Index

9

A

B

I

J

L

M

N

O

P

R